T0329537

Fast-Charging Infrastructure for Electric and Hybrid Electric Vehicles

IEEE Press
445 Hoes Lane
Piscataway, NJ 08854

Fast-Charging Infrastructure for Electric and Hybrid Electric Vehicles

Methods for Large-Scale Penetration into Electric Distribution Networks

Sivaraman Palanisamy
Senior Power Systems Engineer
Vysus Consulting India Pvt Ltd, India

Sharmeela Chenniappan
Professor, Anna University, Chennai, India

Sanjeevikumar Padmanaban
Professor, University of South-Eastern Norway, Norway

Published by John Wiley & Sons, Inc., Hoboken, New Jersey.

Published simultaneously in Canada.

Library of Congress Cataloging-in-Publication Data

Names: Palanisamy, Sivaraman, 1991– author. | Chenniappan, Sharmeela, 1977– author. | Sanjeevikumar, Padmanaban, 1978– author.
Title: Fast-charging infrastructure for electric and hybrid electric vehicles : methods for large scale penetration into electric distribution networks / Sivaraman Palanisamy, Vellalur, Madurai, Tamilnadu, Sharmeela Chenniappan, Adambakkam, Chennai, Tamilnadu, Sanjeevikumar Padmanaban, KPRIET, Anna University Govinda Salai, Pondicherry.
Description: First edition. | Hoboken, New Jersey : Wiley-IEEE Press, [2023] | Includes index.
Identifiers: LCCN 2023008826 (print) | LCCN 2023008827 (ebook) | ISBN 9781119987741 (cloth) | ISBN 9781119987758 (adobe pdf) | ISBN 9781119987765 (epub)
Subjects: LCSH: Battery charging stations (Electric vehicles) | Electric vehicles–Power supply. | Electric vehicles–Electric equipment.
Classification: LCC TL220.5 .P35 2023 (print) | LCC TL220.5 (ebook) | DDC 629.28/6–dc23/eng/20230322
LC record available at https://lccn.loc.gov/2023008826
LC ebook record available at https://lccn.loc.gov/2023008827

Cover Design: Wiley
Cover Image: © Lus Costa/Getty Images

Set in 9.5/12.5pt STIXTwoText by Straive, Pondicherry, India

Dedication

Mr. P. Sivaraman
He is dedicating this book to his father Mr. A. Palanisamy (deceased), mother Mrs. P. Valarmathi, sister Mrs. P. Shanmuga Priya, spouse Mrs. A. Gowri, daughter S. Aathira, and son S. Magizhan.

Dr. C. Sharmeela
She expresses her heartfelt thanks and dedicates this book to her beloved father Mr. N.S. Chenniappan (deceased), and her mother Mrs. C. Kasturi, her brother Sekar, sister-in-law Vanitha, niece Shakthi, friends A. Subathra and M.R. Swaminathan, and all her other beloved friends for encouraging her and extending their full support in writing the book.

Dr. P. Sanjeevikumar
Finally, he also likes to express his sincere thanks to his family members, wife, son, and daughter, for the time sacrificed for his professional activities and the consistent support extended. He also dedicates this book to his father and mother, who provided this opportunity to become an engineer/researcher – their Spirit always Bless him ahead in his Life.

Contents

Preface

Electric Vehicles (EVs) are an absolute necessity due to the rapid depletion of petroleum products, increased operational costs, and environmental pollution. The major barrier to increasing the share of EVs in the market is the charging time. EV charging takes a few minutes to a few hours to recharge the battery completely, and EV users have to wait during the charging process. Compared with AC slow chargers, DC fast chargers can charge EVs in less time. Hence, fast-charging stations or infrastructure is required to increase EV penetration.

This book (Fast-Charging Infrastructure for Electric and Hybrid Electric Vehicles: Methods for Large-Scale Penetration into Electric Distribution Networks) covers the different aspects of fast-charging stations, such as introduction to different types of EV chargers and cables, selection of fast-charging stations, business models and tariff structures, batteries, distribution system planning by DISCOMs, electric distribution infrastructure within the charging stations, energy storage systems, surge protection devices, power quality problems, and standards.

Chapter 1 covers the various equipment/components involved in the fast-charging stations, like cables, chargers, switchgear, distribution transformers, energy meters and power quality meters, plugs, and connectors. Also, it covers renewable energy-powered fast-charging stations and digital communication between the charger and vehicles.

Chapter 2 describes the business model for fast-charging stations, the selection of locations for fast-charging stations, geospatial analysis, and land and electric power supply availability.

Chapter 3 gives detailed information about the integrated and independent business model, the selection of the business model for fast-charging stations, and the tariff structure. Generally, fast-charging stations have two different tariff structures, i.e. 1. tariff between electric utilities (DISCOMs) and fast-charging stations, 2. tariff between fast-charging stations and EV users.

Chapter 4 covers battery chemistries suitable for fast charging, such as the Li-ion family, Lead acid and Nickel family. Also, it covers the selection of battery chemistry for EVs and C-rating.

Chapter 5 describes the need for distribution system planning in terms of electric distribution infrastructure (i.e. distribution feeders and equipment) and power and energy demand requirements to cater for the EV charging loads connected across the distribution system. Real-time examples are given for this purpose.

Chapter 6 provides detailed information about the electric distribution infrastructure within the fast-charging station in terms of reliability. It covers the single point of failure, various redundancy configurations for distribution infrastructure, and chargers.

Chapter 7 covers the requirement of an energy storage system for fast-charging stations and its advantages in combination with renewable energy systems. Also, it covers the various configurations for a renewable energy system and microgrids with and without energy storage systems.

Chapter 8 describes the requirements for surge protective devices for fast-charging stations in an open location, covered location, and underground location.

Chapter 9 gives the power quality problems associated with fast-charging infrastructure like harmonics, transients, voltage unbalance, voltage fluctuations, voltage sags and swells, etc. In detail, it covers the analysis of harmonic injection into the distribution system and the method of analysis, resonance, and super-harmonics.

Standards play an important role in the trouble-free operation of fast-charging stations. Chapter 10 covers various international and national standards for fast-charging stations/infrastructure from an overall perspective, like IEC in Europe, IEEE and SAE in the USA, British standards in the UK, ISO standards, and CEA standards for India.

Chapter 11 provides the requirements for future needs like voltage range, battery chemistry, standardization of battery ratings, distributed energy storage system and grid-friendly charging, ultra-fast chargers of more than 400 kW, interoperable features and wireless charging.

Power electronic converters are used for fast charging purposes. Chapter 12 covers the improved structure of power electronic converters for fast charging.

About the Authors

Mr. P. Sivaraman (Member '20, Senior Member '21 IEEE) was born in Vellalur, Madurai district, Tamilnadu, India. He completed schooling in Govt. Higher Secondary School, Vellalur, B.E. in Electrical and Electronics Engineering, M.E. in Power Systems Engineering from Anna University, Chennai, India in 2012 & 2014, respectively. He has more than eight years of industrial experience in the field of power system studies, renewable energy integration studies, solar PV systems, wind power plant, power quality studies & harmonic assessments, trouble shooting for various power quality problems, providing the techno-economical solution to various power quality problems. Presently he is working as a Senior Power Systems Engineer at Vysus Consulting India Pvt Ltd, India. He has trained more than 500 personnel on renewable energy and power quality. He is an expert in power system simulation software's like ETAP, PSCAD, DIGSILENT POWER FACTORY, PSSE, and MATLAB. He is an active participant in the IEEE standards association. He is a working group member of IEEE standard P2418.5 (Standard for Blockchain in Energy), P1854 (Guide for Smart Distribution Systems), IEEE P2800.2 – Recommended Practice for Test and Verification Procedures for Inverter Based Resources (IBRs) Interconnecting with Bulk Power Systems, IEEE P2844 – Recommended Practice for Limiting Voltage Imbalance in Electric Power Systems and P3001.9 (Design of Power Systems Supplying Lighting Systems in Commercial and Industrial Facilities). He is a working group member of IEEE PES task force of Energy storage. He had authored/co-authored/edited seven books in the field of electrical engineering with Elsevier and Wiley-IEEE Press, published several papers in national and international conferences. He is a senior member of the Institute of Electrical and Electronics Engineers (IEEE), a member of the International Council on Large Electric Systems (CIGRE), a Life Member of the Institution of Engineers (India), and The European Energy Center (EEC). He received Professional Engineer (PEng) certification from the Institution of Engineers India. Also, a speaker who is well versed in both National and International Standards.

Google Scholar link: https://scholar.google.co.in/citations?user=XLdd0mgAAAAJ&hl=en&authuser=1

Dr. C. Sharmeela holds a B.E. in Electrical and Electronics Engineering, M.E. in Power Systems Engineering from Annamalai University, Chidambaram and a

Ph.D. in Electrical Engineering from College of Engineering, Guindy, Anna University, Chennai respectively. At present, she holds the post of Professor and Professor-In-Charge, Power Engineering and Management, Department of Electrical and Electronics Engineering, C.E.G., Anna University, Chennai. She has done a number of consultancies on Renewable Energy Systems such as Solar Photo Voltaic (SPV) Power System, Power quality measurements and design of compensators for industries. She has coordinated and organized several short-term courses on power quality for Tamil Nadu State Electricity Board Engineers, TN, India. She has also delivered several invited talks and trained 1000+ engineers on the importance of Power Quality, Power Quality Standards and Design of SPV power system for more than 12 years in leading organizations such as CII, FICCI, CPRI, MSME, GE (Alsthom) and APQI. She has authored over 30 journal papers in refereed international journals, co-authored 15 book chapters, edited five books and authored one book. Her areas of interest include Power Quality, Power Electronics applications to Power Systems, Smart Grid, Energy Storage Systems, Renewable Energy Systems, Electric Vehicle, Battery Management System and Electric Vehicle Supply Equipment. She is a senior member of IEEE, Fellow of the Institution of Engineers (India), Life Member of ISTE, Central Board of Irrigation and Power (CBIP), New Delhi, India and SSI, India. She has a teaching/research and consultancy experience of 21+ years in the areas of power quality and power systems.

Google scholar link: https://scholar.google.co.in/citations?user=-YbjPxsAAAAJ&hl=en&authuser=1

Dr. Sanjeevikumar Padmanaban (Member' 12, Senior Member' 15 IEEE) received the bachelor's degree from the University of Madras, India, in 2002, the master's degree (Hons.) from Pondicherry University, India, in 2006, and the Ph.D. degree University of Bologna, Italy, in 2012. He was an Associate with various institutions like VIT University India, National Institute of Technology, India, Qatar University, Qatar, Dublin Institute of Technology, Ireland, University of Johannesburg, South Africa. Currently, he is working as a Professor, Department of Electrical Engineering, Information Technology, and Cybernetics, University of South-Eastern Norway, Norway. He has authored 300 plus scientific papers and has received the Best Paper cum Most Excellence Research Paper Award from IET-SEISCON'13, IET-CEAT'16, and five best paper awards from ETAEERE'16 sponsored Lecture note in Electrical Engineering, Springer book series. He is a fellow the Institution of Engineers, FIE, India, fellow the Institution of Telecommunication and Electronics Engineers, FIETE, India, and fellow the Institution of Engineering and Technology, IET, UK. He serves as an Editor/Associate Editor/Editorial Board of the refereed journal, in particular, the IEEE Systems Journal, the IEEE Access Journal, the IET Power Electronics, Journal of Power Electronics, Korea, and the subject editor of the subject Editor of IET Renewable Power Generation, the subject Editor of IET Generation, Transmission and Distribution, and the subject editor of FACTS journal, Canada.

Google Scholar link: https://scholar.google.co.in/citations?user=KyuMg7IAAAAJ&hl=en&authuser=1

Acknowledgments

We thank the almighty for giving us enough strength and support to complete the book.

Mr. P. Sivaraman expresses his sincere thanks to Mr. Balaji Sriram, Research Scholar, IIT Kanpur; D. Sathiya Moorty, Research Scholar, IIT Ropar; Mr. Upendran, Research Scholar, IIT Madras; Mr. S. Rajkumar, Executive, JLL, Bengaluru; Mr. K. Sasikumar, Electrical Engineer, Mott MacDonald, Bengaluru; Mr. Muthukumaran, Director, TECH Engineering Services, Chennai; and Mr. K. Balaji, Electrical Engineer, Vertiv, Chennai, for providing their technical support, figures, expert review, and finalizing the contents.

Dr. C. Sharmeela expresses her sincere gratitude to her mentor, Prof. Dr. D.P. Kothari, and research supervisor, Prof. Dr. M.R. Mohan, Anna University, Chennai. She also takes this opportunity to thank the funding agency, All India Council for Technical Education (AICTE) for funding the research project under the Research Promotion Scheme (RPS) titled "Smart Electric Vehicle Charging Station" in supporting her Electric Vehicle Charging research endeavors. She also likes to express her heartfelt gratitude and thanks to the RUSA 2.0 project – "Electric Vehicle Technologies – Smart Material Characterization, Manufacturing and Grid Management" – Thematic Title "Monitoring and Analysis of Power Quality Issues on to the Distribution Network Due to Electric Vehicle Infrastructure," for providing the test facilities to conduct the Power Quality (PQ) research and analysis in Electric Vehicle Supply Equipment (EVSE).

Dr. P. Sanjeevikumar takes this opportunity to thank the University of South-Eastern Norway, Porsgrunn, Norway, for providing substantial time and facilities to execute his professional activities; the Department of Electrical Engineering, IT and Cybernetics colleagues; his research collaborates/researchers; and the contributed authors for their vital time to make this book a successful outcome.

1

Introduction to Electric Vehicle Fast-Charging Infrastructure

1.1 Introduction

The electric vehicles (EVs) are an absolute necessity due to the rapid degradation of fossil fuels and they are free from environmental pollution during their operation. In a conventional vehicle, petroleum products like petrol or diesel, or gasoline are used as a fuel for transportation. Similarly, in EVs, energy stored in the batteries is used as fuel for transportation. So, batteries in the EVs are to be recharged whenever the battery state of charge (SoC) indication is low or on a need basis by a charger or charging equipment. The charging equipment used to recharge the batteries in EVs is called as electric vehicle supply equipment (EVSE) [1, 2]. An EVSE is placed in a charging station, and it receives the electric power supply from DC or AC supply system and supplies the DC power to recharge the renewable energy storage system (RESS) or simply batteries in the EVs. The typical single-line diagram (SLD) of a charging station is shown in Figure 1.1.

The IEC 61851-1:2017 [3] classifies the EVSE in a charging station into the following types:

1) Characteristics of power supply input
   The EVSE is classified based on the grid supply system used to power the EVSE,
   - EVSE connected to AC grid power supply system
   - EVSE connected to DC grid power supply system
2) Characteristics of power supply output
   The EVSE is classified based on the type of current it will deliver to the EV,
   - AC EVSE
   - DC EVSE
   - AC and DC EVSE

Fast-Charging Infrastructure for Electric and Hybrid Electric Vehicles: Methods for Large-Scale Penetration into Electric Distribution Networks, First Edition. Sivaraman Palanisamy, Sharmeela Chenniappan, and Sanjeevikumar Padmanaban.

Published 2023 by John Wiley & Sons, Inc.

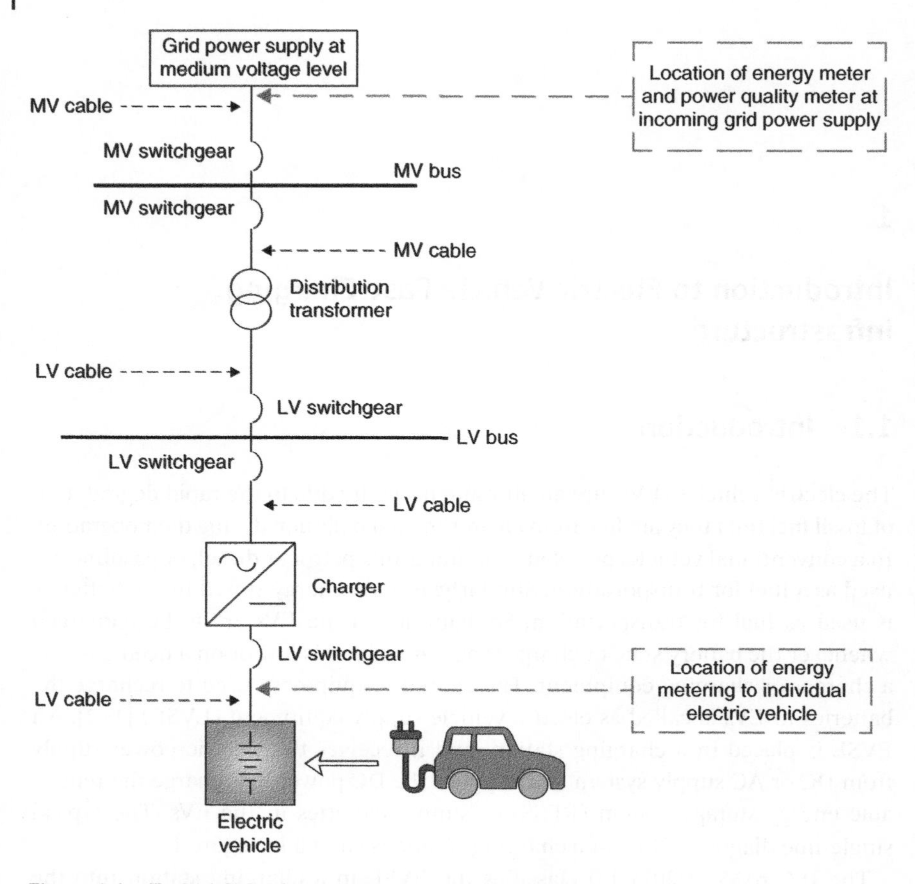

Figure 1.1 Typical SLD of a charging station. MV, medium voltage.

3) Based on the type of electric connection method
   - Plug and cable connected
   - Permanently connected
4) Based on environmental conditions
   - Indoor
   - Outdoor
5) Based on access to the EV users
   - Locations with restricted access
   - Location with non-restricted access
6) Based on the mounting method
   - Stationary EVSE equipment
   - Non-stationary EVSE equipment

7) Based on protection against electric shock
   - Class I equipment
   - Class II equipment
8) Based on charging modes
   - Mode 1
   - Mode 2
   - Mode 3
   - Mode 4

These charging equipment are generally classified into AC charging and DC charging based on where the actual conversion of AC to DC takes place. In AC charging method, the actual conversion of AC to DC takes place inside the EV or onboard. The different types of AC chargers available in the markets are listed in Table 1.1.

In the DC charging method, the actual AC to DC conversion takes place outside of EV (i.e. inside of EVSE) or off-board. The different types of DC chargers available in the markets are listed in Table 1.2.

Also, based on the time taken to recharge the battery to 100% SoC, EVSE is classified into slow charging and fast charging. In the slow charging method, it will take six to eight hours or above to recharge the battery to 100% SoC. This slow charging method is also called overnight charging. In the fast-charging method, EVs are recharged at a higher power rating charger and usually, it will take

Table 1.1 List of AC chargers.

S. No	Charger type	Connector/Socket	Maximum power output
1	Type 1	Yazaki socket	Up to 7.4 kW (32 A, single phase)
2	Type 2	Mennekes socket	Up to 44 kW (63 A, three Phase)
3	Type 3	Grand socket	Up to 22 kW (32 A, three Phase)

Table 1.2 List of DC chargers.

S. No	Charger type	Maximum power output	Communication protocols
1	CHAdeMO	Up to 400 kW DC charging (1000 V, 400 A)	Control Area Network (CAN) for communication between EV and EVSE
2	GB/T	Up to 237.5 kW DC charging (950 V × 250 A)	CAN for communication between EV and EVSE
3	Tesla super charger	Up to 135 kW DC charging (410 V × 330 A)	CAN for communication between EV and EVSE

15–30 minutes to recharge the battery up to 80% SoC [2]. So, the fast-charging method is a widely adopted method to recharge EVs used for commercial purposes including freight transfers. Based on the location of the charging process carried out, charging stations are also classified into residential/home charging and public charging. Residential charging is widely used by EVs owner because recharging is economical at home (i.e. energy cost is less). It normally uses slow chargers to recharge the battery and EVs owner charge their vehicle mostly during the night-time [4]. On the other hand, public charging stations (e.g. charging station at shopping mall, cinema theater, highway, etc.) allow EV users to recharge their vehicle outside the residential premises. The public charging station generally employs DC fast chargers to recharge EV batteries in lesser time to avoid the waiting period.

This chapter discusses the fast-charging infrastructure as a public charging station to recharge an electric vehicle.

1.2 Fast-Charging Station

A charging station is like a fuel stop or a petrol station situated in most of the public places in and around the city and highways. It has off-board chargers of high-power rating and is used to recharge the EV batteries with DC current in lesser time duration. The advancements in battery technologies that allow higher charging rates and higher charging cycles make the fast charging a realistic possibility [5–7]. The SLD of typical fast-charging station is shown in Figure 1.2.

The main elements of any fast-charging station are

1) Power grid or grid power supply
2) Power cables
3) Switchgear
4) Transformer
5) Energy meters and power quality meters
6) Fast chargers
7) Plugs and connectors

1.2.1 Power Grid or Grid Power Supply

The power grid is the main source of power supply to the connected loads (including EV loads and other auxiliary loads) of a fast-charging station. Generally, the grid power supply is provided by the distribution company (DISCOMs) in medium voltage level (e.g. 11, 22, or 33 kV based on the particular country voltage level). Some places, where the charging station loads are less, DISCOMs provides the grid power

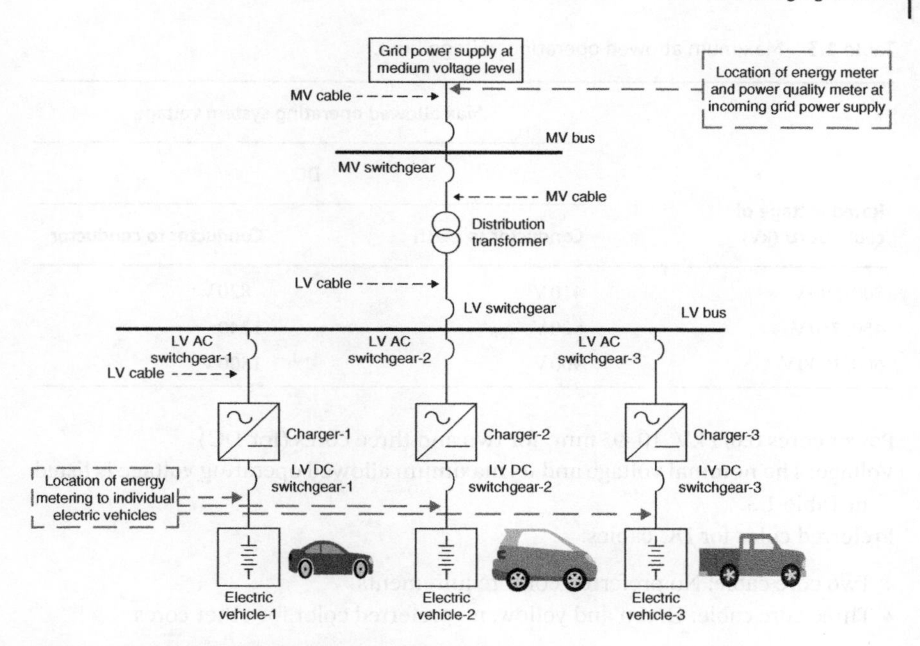

Figure 1.2 Typical SLD of a fast-charging station.

supply at low voltage level (e.g. 400 or 415 V) and it does not require the dedicated transformer powered by grid power supply at the medium voltage level.

1.2.2 Power Cables

The power cables are used to carry the electric power from one place to another place. In fast-charging station, power cables are used between the power grid supply source to distribution transformer (medium voltage side), distribution transformer to low voltage switchgear (low voltage side), low voltage switchgear to fast charger (input AC side), and fast charger (output terminal) to EV batteries (DC cables).

The power cables between the EVSE (fast charger DC output terminal) and EV shall comply the technical requirements of IEC 62893 Charging cables for electric vehicles for rated voltages up to and including 0.6/1 kV including the test requirements/methods.

IEC 62893-1 gives the general requirements of the cables for Mode 1–3 (AC charging) [8] and Mode 4 (DC charging). Some of the requirements of IEC 62893-1 related to DC charging are as follows [1, 9]:

Conductor: Plain or tinned annealed copper comply with Class 5 of IEC 60228
Number of cores: Two core cable and three core cable

Table 1.3 Maximum allowed operating voltage limit.

Rated voltage of cable Uo/U (kV)	Max allowed operating system voltage	
	DC	
	Conductor to earth	Conductor to conductor
300/500 V	410 V	820 V
450/750 V	620 V	1240
600/1000 V	900 V	1800 V

Power cores 0.6/1 kV: 10–95 mm^2 for two and three cores (for DC)
Voltage: The nominal voltage and its maximum allowed operating voltage is listed in Table 1.3.
Preferred color for DC cables:

- Two core cable: No preferred color requirements
- Three core cable: Green and yellow, no preferred color for other cores

Insulation material:

- Power cores: Compound of EVI-2
- Pilot, control, and any other additional cores: EVI-1 or EVI-2

Metallic screen (optional): Braid of plain or tinned copper wire
Sheath: Compound of EVM-1, EVM-2, or EVM-3

IEC 62893-4-1 gives the technical requirements of the cables for mode 4 DC charging without use of a thermal management system are as follows [10]:

Rated voltage: 1.5 kV DC
Power cores: 4–150 mm^2 for two or more cores
Auxiliary power cores (optional): 2.5–6 mm^2 – two cores
Temperature sensor cores (optional): Minimum size is 0.5 mm^2
Conductor: Plain or tinned annealed copper comply with Class 5 of IEC 60228
Number of cores: Two core cable and three core cable
Preferred color for DC cables:

- Two core cable: No preferred color requirements
- Three core cable: Green and yellow, no preferred color for other cores

Insulation material:

- Power cores: Compound of EVI-2
- Pilot, control, and any other additional cores: EVI-1 or EVI-2

Table 1.4 Recommended use of cables without thermal management for EV charging.

S. No	Code designation	Cable type	Recommended use	Application	Remarks
1	62893 IEC 126	EVM-1	Mode 4 as per IEC 61851-1	Indoor and outdoor	—
2	62893 IEC 127	EVM-2			
3	62893 IEC 128	EVM-3		Outdoor	Used only for outdoor applications because it will release the halogen content during fire

Metallic screen (optional): Braid of plain or tinned copper wire
Sheath: Compound of EVM-1, EVM-2, or EVM-3

The recommended use of cables without thermal management for EV charging is listed in Table 1.4.

IEC 62893-4-2 gives the technical requirements of the cables for mode 4 DC charging with thermal management system are as follows [11]:

Rated voltage: 1.5 kV DC
Power cores: 16–150 mm^2 for two or more cores
Auxiliary power cores (optional): 2.5–6 mm^2 – two cores
Temperature sensor cores (optional): Minimum size is 0.5 mm^2
Conductor: Plain or tinned annealed copper comply with Class 5 of IEC 60228
Number of cores: Two core cable and three core cable
Preferred color for DC cables:

- Two core cable: No preferred color requirements
- Three core cable: Green and yellow, no preferred color for other cores

Insulation material:

- Power cores: Compound of EVI-2
- Pilot, control, temperature sensor, and auxiliary power cores: EVI-1 or EVI-2

Metallic screen (optional): Braid copper with minimum 80% optical coverage
Tubes: Fluid-filled tubes should be made resistant to the media used as a coolant. It has to withstand an aging of 120 °C temperature for 168 hours in dry air. If the tube is also used as an insulator and it has to comply the requirements of EVI-2.
Sheath: Compound of EVM-1, EVM-2, or EVM-3
Liquid coolants: The coolant type used in the cable shall be specified by the manufacturer of the system. It has to satisfy the test requirements of IEC 62893-4-2.

Table 1.5 Recommended use of cables with thermal management for EV charging.

S. No	Code designation	Cable type	Recommended use	Application	Limitations	Remarks
1	62893 IEC 129	EVM-1	Mode 4 as per IEC 61851-1 with thermal management system	Indoor and outdoor	During the normal operating conditions, the cable maximum surface voltage should not exceed 60°C	—
2	62893 IEC 130	EVM-2				
3	62893 IEC 131	EVM-3	Mode 4 as per IEC 61851-1 with thermal management system	Outdoor		Used only for outdoor applications because it will release the halogen content during fire

The recommended use of cables with thermal management for EV charging is listed in Table 1.5.

1.2.3 Switchgears

The switchgears are the protection and control equipment used to make and break the circuit by means of automatic or manual operation in the event of electric faults and/or any maintenance activities. Generally, in fast-charging stations, switchgears are used at medium voltage level (i.e. distribution transformer high voltage side), low voltage level (i.e. distribution transformer low voltage side, individual outgoing feeders to fast chargers and auxiliary loads), and output of fast chargers (DC supply) to the EV.

Each EVSE to EV circuit/feeder is protected by a circuit breaker complying IEC 60947 and a residual current device (RCD) of type A or type B for a rated residual operating current not more than 30 mA, as required by IEC 60364-7-722.

1.2.4 Distribution Transformer

A distribution transformer is an electromechanical device that is used to step down the voltage without changing the frequency. In a fast-charging station, distribution transformers are used to transfer the electric power at a medium voltage level (e.g. 11 kV) to a low voltage level (e.g. 415 V). The size/rating of the distribution transformer should be selected based on the respective fast-charging station power demand. The number of distribution transformer requirements depends on the reliability aspects of the charging station. Refer to Chapter 6 for more details.

Figure 1.3 Typical distribution transformer.

The typical distribution transformer is shown in Figure 1.3.

1.2.5 Energy Meters and Power Quality Meters

The energy meters are used to measure the energy consumed by the charging station from DISCOMs. Generally, this meter will be installed at the interconnection point (i.e. point of common coupling – PCC). The actual billing for energy consumed by the charging station is based on this meter reading. The technical requirement for this meter will depend on the country grid code (transmission code or distribution code). Generally, an energy meter with 0.2S class accuracy is used for billing purposes.

Fast chargers in the charging station are used to recharge the batteries in the EV. It means, energy is transferred from the charging station to the batteries in the EV. The EV user has to pay for the actual energy that is used to recharge the batteries. Hence, one more set of energy meters installed at the output side of the fast charger or an inbuilt feature of energy metering in the fast charger is used for billing purposes.

The power quality meters are used to measure the various power quality characteristics (like voltage sag, voltage swell, interruption, transients, harmonics, flicker, rapid voltage change, etc.) of the fast-charging station during their operation. Generally, this power quality meter will be installed at PCC. The charging stations are receiving the grid power supply from DISCOMs and it has to maintain the grid code requirements (including power quality requirements) of the country. This power quality meter is used to check the compliance of the charging station with respect to grid code power quality requirements. As per IEC 61000-4-30, the technical requirements of a power quality meter for compliance checking/verification shall be Class A.

Figure 1.4 Typical fast charger.

1.2.6 Fast Chargers

The fast chargers are used to recharge the EV batteries in a lesser time duration. It converts AC into DC (rectifier) and increases or decreases the output DC voltage to recharge batteries (as a DC–DC buck/boost converter). The power rating of DC fast chargers is typically in the range of 50–150 kW for light motor vehicles and more than 150 kW for electric buses and heavy motor vehicles.

The typical fast charger is shown in Figure 1.4.

1.2.7 Plugs and Connectors

The plug and connectors are used to connect the EV to the fast charger, i.e. it establishes the temporary connection between the EVSE and EV during the charging period. The IEC 61851-1 describes the three different methods of connection between the EV and EVSE [3].

Method 1: The connection between the EV and EVSE with a cable and plug attached permanently to the EV as shown in Figure 1.5 [3].

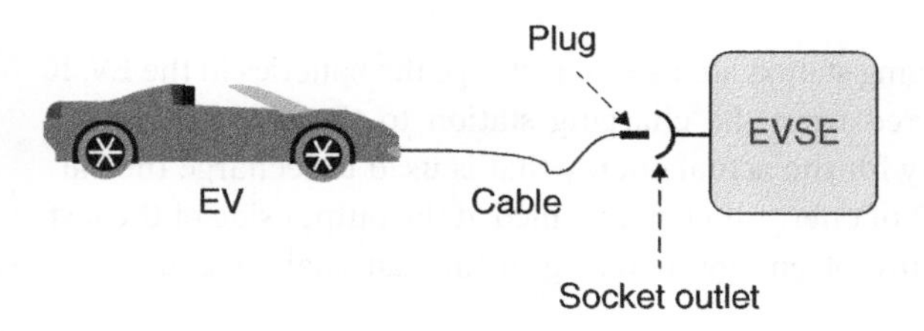

Figure 1.5 Connection between the EV and EVSE with a cable and plug attached permanently to EV.

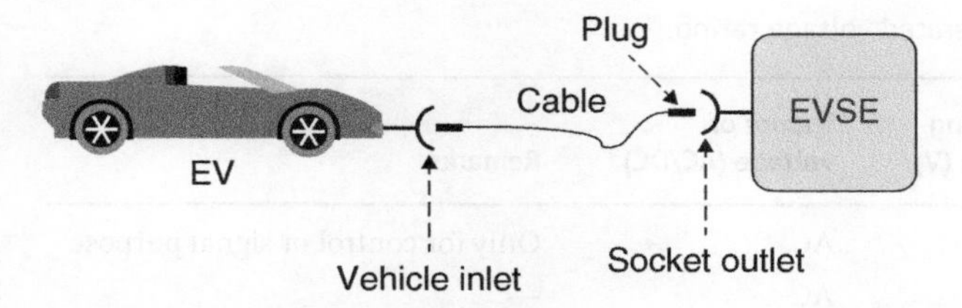

Figure 1.6 Connection between the EV and EVSE with detachable/removable cable assembly at both EV and EVSE end.

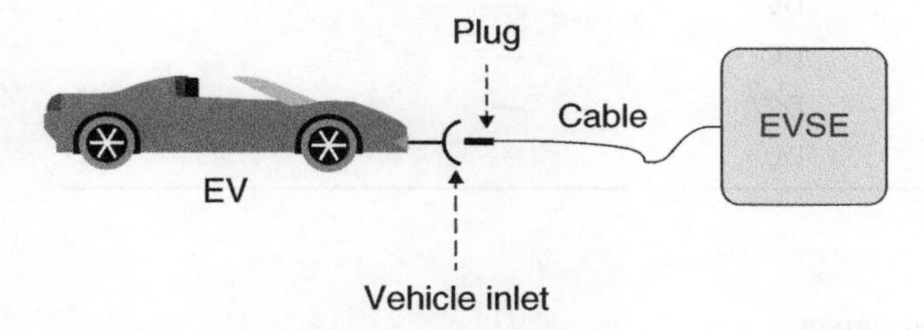

Figure 1.7 Connection between the EV and EVSE with a cable and plug attached permanently with EVSE.

Method 2: The connection between the EV and EVSE with detachable/removable cable assembly at both EV and EVSE end as shown in Figure 1.6 [3].

Method 3: The connection between the EV and EVSE with a cable and plug attached permanently with EVSE as shown in Figure 1.7 [3].

In the mated position (i.e. connection established between the plug and socket outlet), the connection shall comply with the minimum requirements of IPXXD as per ISO 20653.

The requirements (electrical and mechanical) for connection between an electric vehicle and EVSE for plugs, socket-outlets, vehicle inlets, and vehicle connectors are covered in IEC 62196 (all parts) [12, 13].

The requirements specified by IEC 62196-3 are applicable to mode 4 charging as per IEC 61851-1, for the rated voltage up to 1500 V DC and current up to 250 A, rated voltage 1000 V AC, and current up to 250 A. The vehicle couplers specified in IEC 62196 are applicable to an ambient temperature from −30°C to +50°C and are intended to connect with copper or copper alloy cable.

The preferred operated voltage rating as per clause 5 of IEC 62196-1 is listed in Table 1.6 [12].

The preferred rated current rating as per clause 5 of IEC 62196-1 is listed in Table 1.7 [12].

Table 1.6 Preferred operated voltage rating.

S. No	Rated operating voltage range (V)	Nature of voltage (AC/DC)	Remarks
1	0–30	AC	Only for control or signal purpose
2	100–130	AC	—
3	200–250	AC	—
4	380–480	AC	—
5	600–690	AC	—
6	480	DC	—
7	600	DC	—
8	750	DC	—
9	1000	DC	—

Table 1.7 Preferred rated current.

S. No	Preferred rated current (A)	Remarks
1	5	—
2	13	—
3	16–20	—
4	30–32	—
5	60–63	—
6	70	—
7	80	Only for DC
8	125	—
9	200	Only for DC
10	250	—
11	400	Only for DC
12	500	Only for DC
13	600	Only for DC
14	630	Only for DC
15	800	Only for DC

The DC charging connector types are generally classified as follows:

1) CCS-combo 1 (used in the United States)
2) CCS-combo 2 (used in Europe)
3) CHAdeMO connector (used globally for cars manufactured by Japanese automakers)
4) Tesla DC connectors (AC as well as DC)
5) DC connector based on the Chinese standard GB/T

The combined charging system (CCS) connectors are also called hybrid connectors because they are built-in connectors for AC and DC charging. In fact, they derive from Type 1 and Type 2 connectors for AC load which are now referred to as Combo 1 and Combo 2, respectively.

It uses the SAE J1772 load input and adds two more pins underneath [14]. It "combines" the SAE J1772 connector with the high-speed charging pin. The CCS is the accepted standard in North America and was developed and approved by the Society of Automotive Engineers (SAE). Most car manufacturers are using the CCS standard in North America, including General Motors, Ford, Chrysler, Dodge, Jeep, BMW, Mercedes, Volkswagen, Audi, Porsche, Honda, Kia, Fiat, Hyundai, Volvo, smart, MINI, Jaguar Land Rover, Bentley, Rolls Royce, and others [15].

1.2.7.1 CCS Combo 1 Connector

The combo 1 vehicle connector is derived from a Type 1 AC connector and retains the ground pin and the two signal pins (pilot control and proximity pilot). Additionally, two additional DC power pins are at the bottom of the connector for fast charging. On the vehicle input, the pin configuration is the same as the type 1 AC connector at the top, while the bottom two pins are used for DC charging.

Similarly, the CCS combo 2 connectors are derived from the Type 2 AC connector, which holds the ground pin and both signal pins (Control pilot and proximity pilot), and the two DC power pins are added to the bottom of the high-power DC charger.

In this type of DC charging, power line communication (PLC) is used in both combo 1 and 2. PLC is a technology that carries data for communication on existing electric power lines which is used simultaneously for power transmission as well.

1.2.7.2 CHAdeMO Connector

CHAdeMO connector is developed by the Japanese utility Tepco and almost all DC fast chargers in Japan used the CHAdeMO connector. Unlike CCS combo 1 and 2, CHAdeMO connectors do not share part of the connector with the

SAE J1772 input, so they require an additional CHAdeMO input on the car. This requires a larger charging port area to accommodate two separate charging sockets [17].

It has three power pins and six signal pins for its operation, and it uses the CAN (controlled area network) protocol for communication between the vehicle and EVSE (charger).

CAN communication is a robust vehicle communication standard designed to allow microcontrollers and devices to communicate with each other in real-time without a host computer.

1.2.7.3 Tesla Connectors

Tesla superchargers in United States use their own connectors, while their European variant uses the Type 2 connector.

1.3 Fast-Charging Station Using Renewable Power Sources (RES)

The EVs are charged through the fast-charging station as public charging infrastructure. These EV charging loads are added additional to the existing distribution system loads in many locations, and it requires the additional distribution network infrastructure. Also, it increases the power as well as energy demand of the distribution system. In order to reduce the power demand due to EV charging and environmental pollution problems, renewable energy sources are locally used to generate the power and integrated with the fast-charging station. In other words, all EVs have a benefit over environmental pollution, i.e. carbon emissions from the EVs are zero. Certainly, there are carbon emissions at some point in the overall energy conversion process whenever fossil fuels are used to generate electricity for recharging the EVs. In order to avoid carbon emissions throughout the EV operation, renewable energy sources like solar PV systems, wind energy conversion systems, hydropower, biomass, or fuel cells are preferred to generate the electricity of EVs.

The integration of renewable-powered fast-charging stations along with an energy storage system (ESS) reduces the power demand of the distribution grid to a greater extent during the large penetration of EV charging stations. Another important advantage of using RES will reduce the operational cost (i.e. cost of energy consumed from DISCOMs). Generally, the RES integration into a fast-charging station can be done in two ways.

1) Integration of RES at AC bus of the fast-charging station.
2) Integration of RES at DC bus of the fast-charging station.

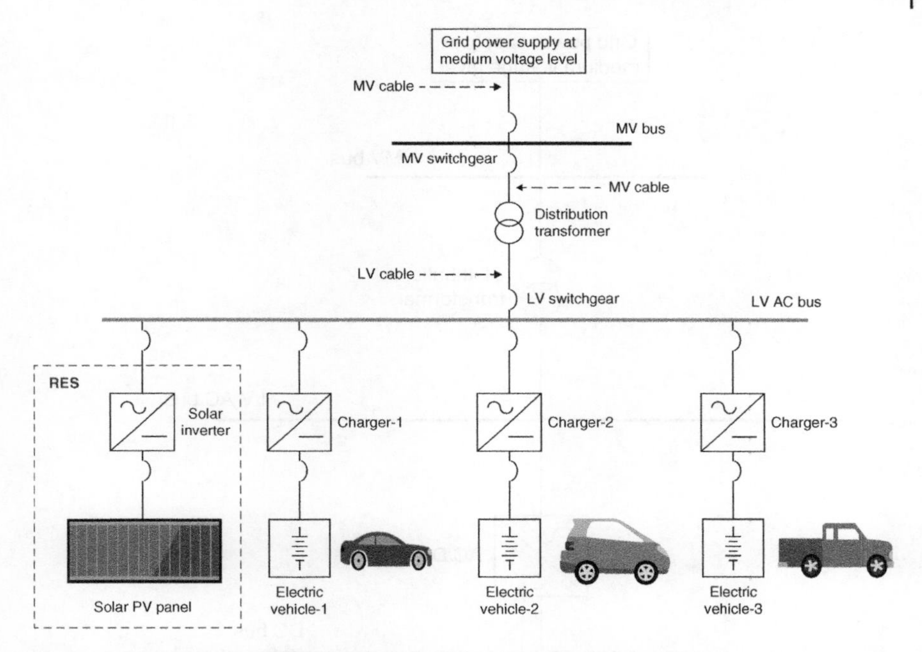

Figure 1.8 Integration of RES at AC bus of the fast-charging station.

The typical Integration of RES at the AC bus of the fast-charging station is shown in Figure 1.8. The solar PV system (one of the RES) is integrated into the LV AC bus through the solar inverter (solar inverter will convert the DC supply into AC supply).

The typical integration of RES at the DC bus of the fast-charging station is shown in Figure 1.9. The electricity generated by solar PV systems and batteries of EV charging is on DC supply. Hence, the integration of RES directly into the DC bus shall reduce the energy loss in DC to AC conversion in solar inverter and AC to DC conversion in EV charger.

The ESS is also used in some fast-charging stations along with RES. It is used to store the energy whenever the solar power generation is higher than EV charging loads or whenever the DISCOMs electricity tariff is less during the off-peak hours. The energy stored in the ESS is used to recharge the EV batteries whenever the solar power generation is lesser than the EV charging loads or whenever the DISCOMs tariff is high during peak hours. The typical Integration of the battery energy storage system (BESS) at the DC bus of the fast-charging station is shown in Figure 1.10.

This configuration shares the common DC bus for all the sources and loads (EVs) in a fast-charging station. The energy management system (EMS) plays an

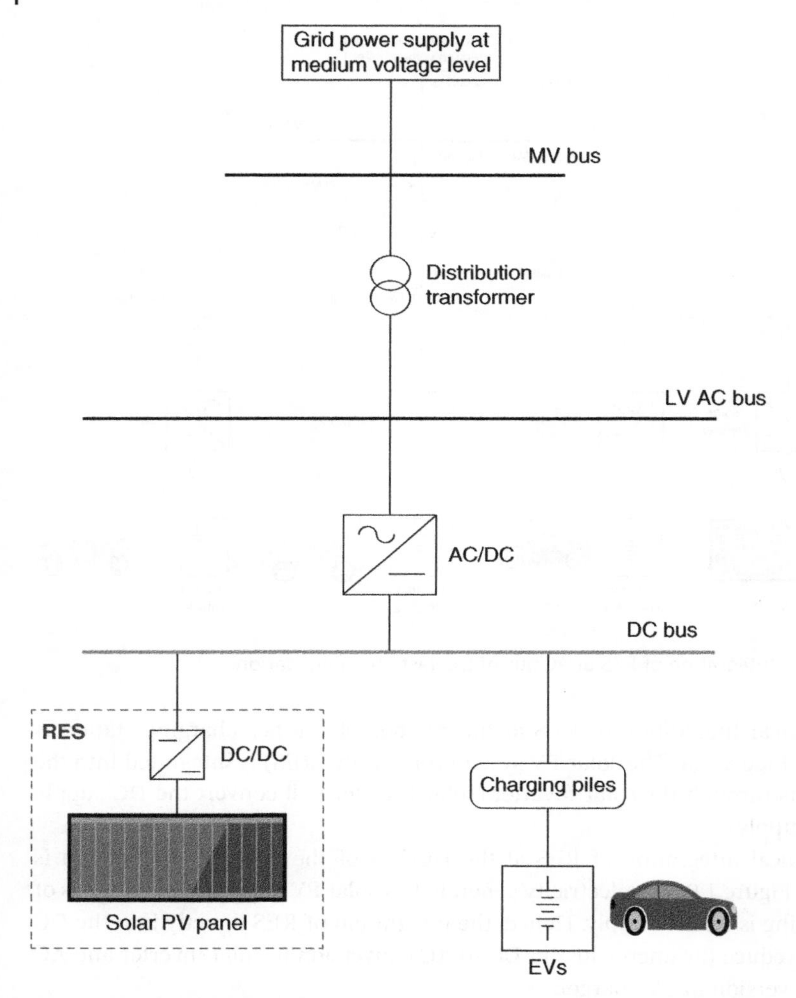

Figure 1.9 Integration of RES at DC bus of the fast-charging station.

important role in optimal power scheduling and power management. The EVs are charged by the power from the solar PV system during the solar generation time (typically 07:00 a.m.–17:00 p.m.). If the solar generation is higher than the EV charging load demand, excessive generation is used to charge the BESS. Whenever the solar generation is lesser than the EV charging load demand, BESS is used to meet the EV charging load demand in addition to the solar power generation. If BESS storage is not sufficient, then the grid power supply is used to meet the EV

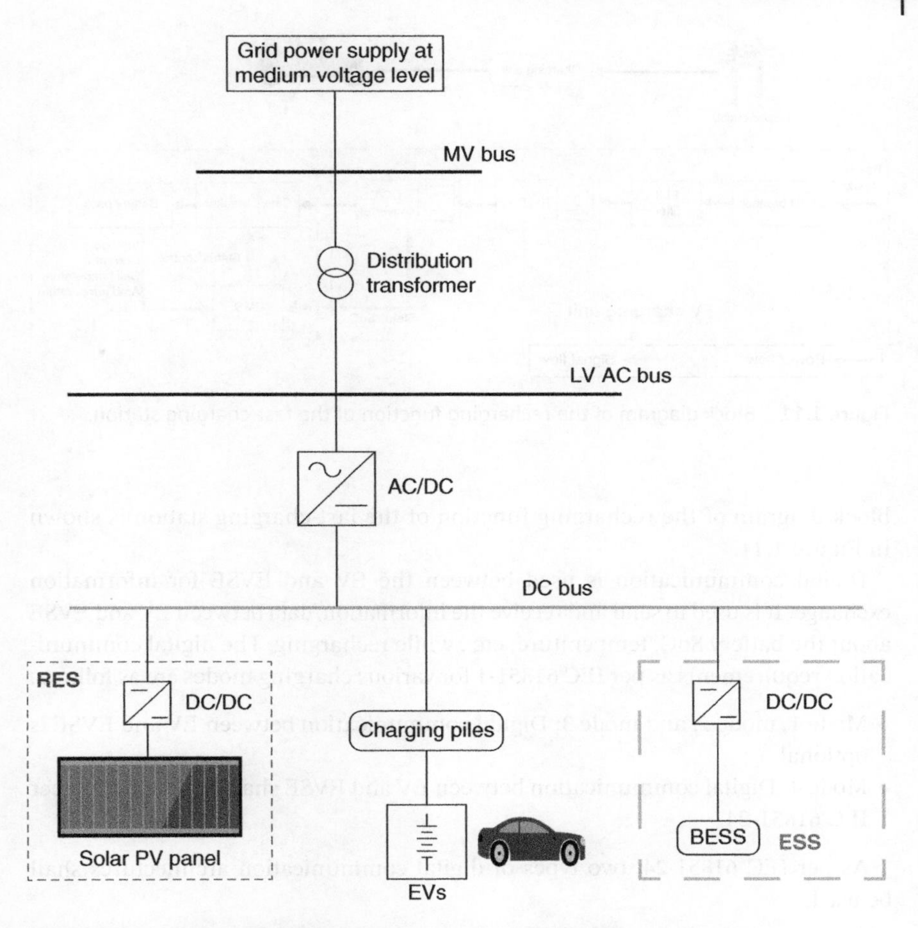

Figure 1.10 Integration of BESS at DC bus of the fast-charging station.

charging loads. The EMS will take care of the entire power and energy management of the fast-charging station.

1.4 Digital Communication for Fast-Charging Station

Digital communication is required between the EV and EVSE for safe and trouble-free recharging. Digital communication is defined as the digitally encoded information exchange between the EV and EVSE. The typical

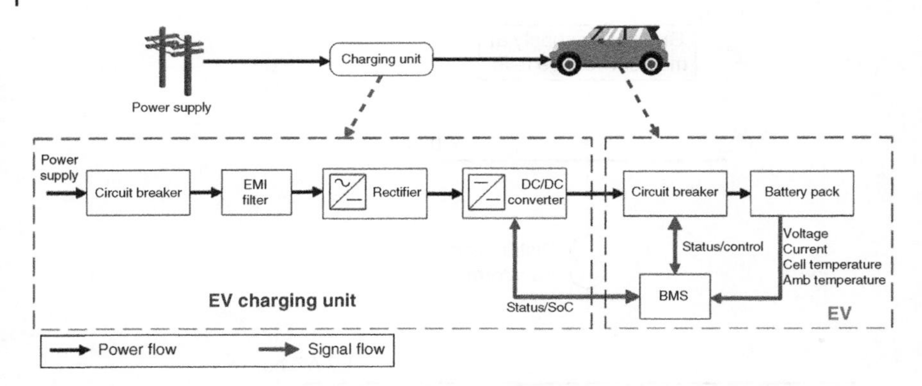

Figure 1.11 Block diagram of the recharging function of the fast-charging station.

block diagram of the recharging function of the fast-charging station is shown in Figure 1.11.

Digital communication is used between the EV and EVSE for information exchange. It is used to send and receive the information/data between EV and EVSE about the battery SoC, temperature, etc., while recharging. The digital communication requirements as per IEC 61851-1 for various charging modes are as follows:

- Mode 1, mode 2, and mode 3: Digital communication between EV and EVSE is optional
- Mode 4: Digital communication between EV and EVSE shall be provided as per IEC 61851-24

As per IEC 61851-24, two types of digital communication architectures shall be used,

- ✓ Control Area Network (CAN): Using the dedicated circuit for data communication using CAN protocol as per ISO 11898-1.
- ✓ Homeplug Green PHY: Using the control pilot line.

Some of the information exchange between the EV and EVSE through digital communication as per IEC 61851-24 is as follows:

- Voltage request for the controlled voltage charging (CVC) system
- The current request for the controlled current charging (CCC) system
- Maximum rated voltage of DC EV charging station
- Maximum rated current of DC EV charging station
- Maximum voltage limit of EV

➢ EV minimum current limit, only for the controlled voltage charging (CVC) system
➢ Zero current confirmed
➢ Charging stopped by the user
➢ EVSE real-time available load current
➢ Loss of digital communication

1.5 Requirements for Fast-Charging Station

In order to perform safe and reliable charging, both electric vehicles and EVSE (chargers) have to meet certain technical requirements as provided by IEEE Std 2030.1.1-2021 [19]. Some of those requirements are listed below.

❖ During the charging period, both the vehicle and EVSE (charger) shall confirm they are physically and/or electrically connected to each other.
❖ The EVSE (charger) shall have the feature to detect circuit isolation, short circuit in the circuits, and ground faults.
❖ The EVSE (charger) shall have the feature of overvoltage protection function.
❖ If the protective conductor is disconnected between the vehicle and EVSE (charger), the charger has to stop charging (charging process) within 10 seconds.
❖ If the communication between the vehicle and EVSE (charger) is interrupted, the charger has to stop the charging process.
❖ The EVSE (charger) shall have the earth leakage current detection and automatic disconnection function to avoid the electric shock.
❖ It has to provide the status of the charging process to users.
❖ When the connector is disconnected or not connected to the vehicle, the EVSE (charger) shall have the feature of not supplying the voltage that is unsafe to the human.
❖ During the charging process, it has to prevent users from touching/accessing the live parts or electrically energized parts.
❖ The EVSE (charger) shall have protection against overload and short circuit of both incoming AC circuit as well as charger internal circuit.
❖ After the disconnection of the vehicle from EVSE, the voltage between any accessible conductive parts (it includes the connector, charging cable, and any grounding parts) is reduced to less than 60 V within one second or less.
❖ A communications interface must be used to transfer/exchange the date between the vehicle and EVSE for charging control.

Table 1.8 Chargers and their ratings.

S. No	Charger type	Charger connectors	Rated voltage (V)	Number of charging points/Number of charging guns
1	Fast charging	CCS (min 50 kW)	200–1000	1/1 CG
		CHAdeMO (min 50 kW)	200–1000	1/1 CG
		Type-2 AC (min 22 kW)	380–480	1/1 CG
2	Slow/Moderate charging	Bharat DC-001 (15 kW)	72–200	1/1 CG
		Bharat AC-001 (10 kW)	230	3/3 CG of 3.3 kW each

In India, any public charging infrastructure must meet the minimum technical requirements stated by the Central Electricity Authority (CEA). They are:

- Every charging station must have an exclusive or dedicated transformer with all related substation equipment including safety appliances.
- It should have 33/11 kV lines/cables and associated equipment along with all the line termination and metering tools, etc.
- It must have adequate space for charging and entry/exit of vehicles.
- It must have one or more electric kiosks/boards with the installation of all the charger models listed in Table 1.8.
- It must be suitable to install the additional number of Kiosks/Chargers in addition to the minimum number of chargers listed in Table 1.8.
- It must be tie up with at least one online network service provider to enable online or advanced remote booking by EV owners for charging slots.
- Appropriate liquid-cooled cables if high-speed charging facility for onboard charging of fluid-cooled batteries.
- Appropriate climate control equipment for fast charging of batteries to be used for swapping.

1.6 Case Study: Public Fast-Charging Station in India

A charge point operator developed the public charging station with a DC fast charger and AC slow charger for public charging. The charging station receives the electric power supply from DISCOM at 11 kV voltage level and uses the 500 kVA, 11 kV/415 V distribution transformer to power the connected loads. The SLD of the charging station is shown in Figure 1.12.

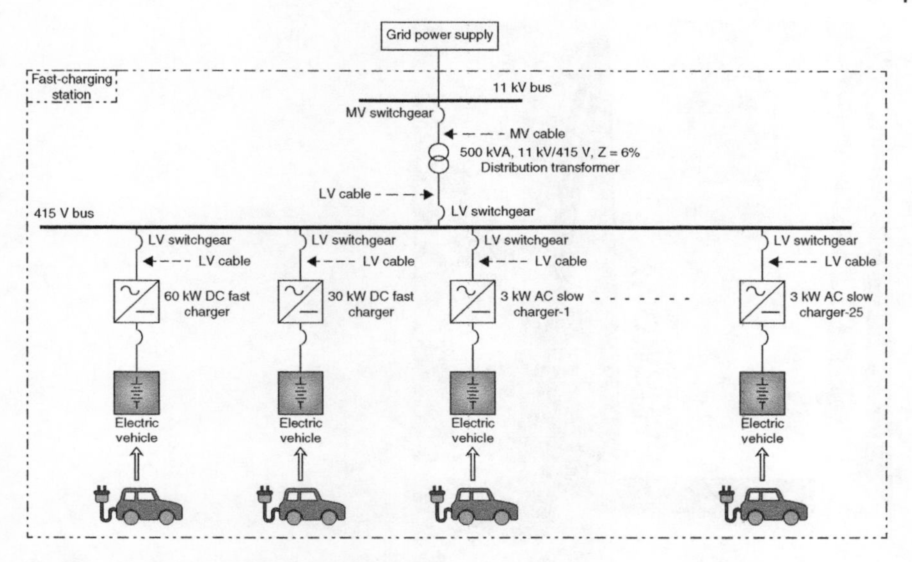

Figure 1.12 SLD of the fast-charging station.

Table 1.9 Chargers used in the fast-charging station.

S. No	Charger type	Power rating (kW)	Number of chargers	Number of charging guns	Installation type	Connection type
1	DC fast charger	60	1	2	Fixed	CCS
2	DC fast charger	30	1	2	Fixed	GB/T
3	AC slow charger	3	25	—	Portable	—

The chargers used in the fast-charging station are listed in Table 1.9.

The chargers (fast charger 60 kW, fast charger 30 kW, and slow charger 3 kW) installation at the charging station is shown in Figures 1.13–1.15, respectively.

The fast charger of 30 kW powering the two electric vehicles (i.e. one charging gun is connected to the first EV and the second charging gun is connected to the second EV) is shown in Figure 1.16.

Whenever any EV is coming for charging, the technician or operator person in the charging station connects the vehicle to the charger. In order to begin the charging process, the technician or operator person in the charging station must initiate the

Figure 1.13 DC fast charger 60 kW. *Source:* Courtesy: EXICOM.

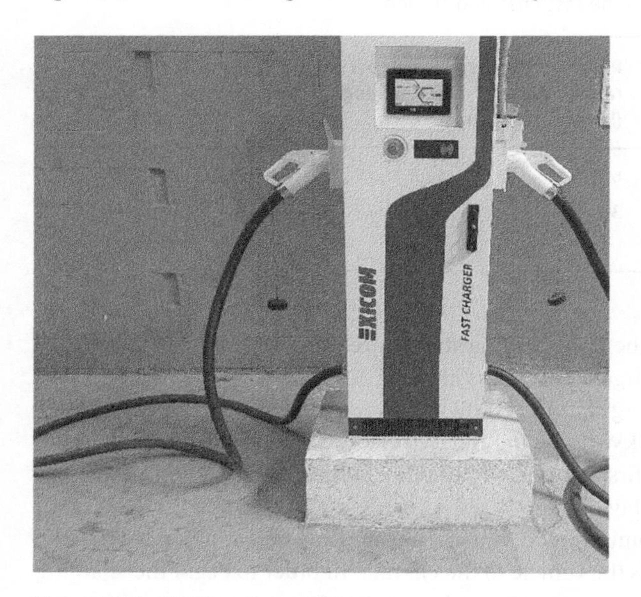

Figure 1.14 DC fast charger 30 kW. *Source:* Courtesy: EXICOM.

charging process (i.e. transferring the energy from the charger to EV) by means of scanning the barcode in the mobile or using the RFID card. This method avoids the unauthorized EVs connected to the charger for charging. After the completion of the charging process, the summary of the charging will be sent to EV users through mobile and users have to pay for the actual energy consumption.

The typical charging parameter of a charger during the charging process in the charger display unit is shown in Figure 1.17.

EV users can use the mobile app for prebooking slots available at the charging station for charging services.

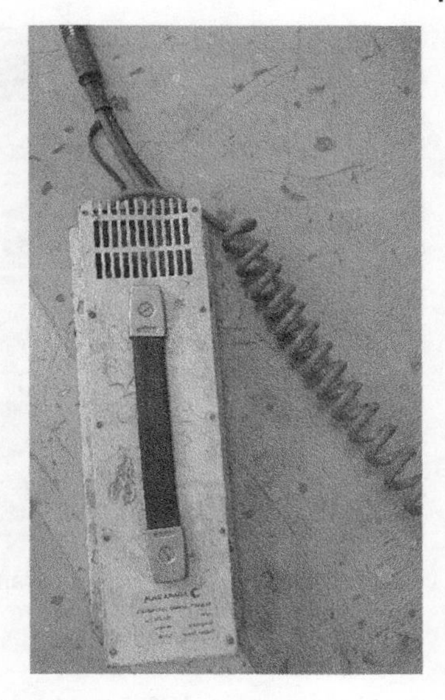

Figure 1.15 AC slow charger 3 kW. *Source:* Courtesy: AMARA RAJA.

1.7 Conclusion

This chapter discusses the various aspects of fast-charging stations like equipment (transformer, charger, connectors, cables, energy meters, etc.),

Figure 1.16 DC fast charger 30 kW charging two EVs at a time.

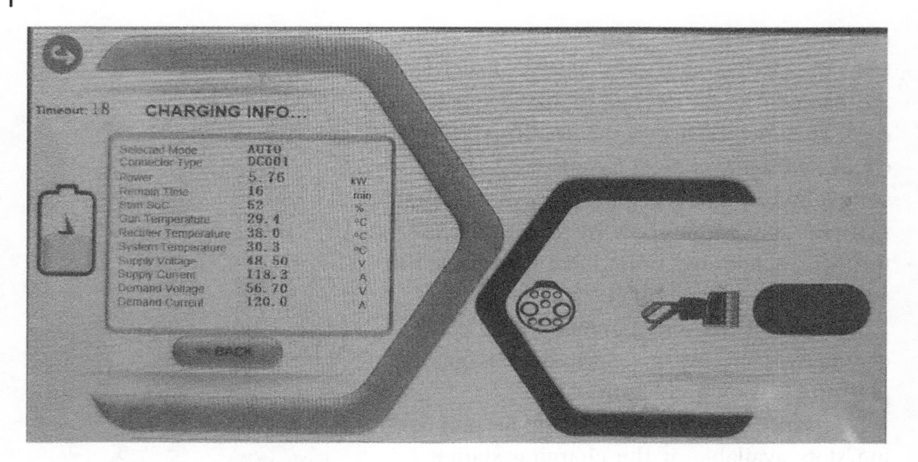

Figure 1.17 Typical charging parameter of a charger during the charging process.

requirements of IEC standards for charging stations, digital communication between vehicle and EVSE, and renewable energy-based fast-charging stations in detail. Also, the requirements of EVSE (fast chargers) for safe and reliable charging as per IEEE 2030.1.1-2021 are discussed.

References

1 Reji Kumar Pillai, Reena Suri, Suddhasatta Kundu, Harpreet Singh, Shuvam Sarkar Roy, and Shreekant Dhuri (2018). Electric vehicle charging stations business models for India, version 1.0, ISGF white paper, pp. 1–15, 1st September 2018.

2 Sivaraman, P. and Sharmeela, C. (2020). Power quality problems associated with electric vehicle charging infrastructure. In: *Power Quality in Modern Power Systems* (ed. P. Sanjeevikumar, C. Sharmeela, and J.B. Holm-Nielsen). London: Academic Press. https://doi.org/10.1016/B978-0-12-823346-7.00005-0.

3 IEC 61851-1:2017 (2017). *Electric vehicle conductive charging system – Part 1: General requirements*. 12-15 IEC Standards.

4 Sivaraman, P., Sharmeela, C., and Logeshkumar, S. (2021). Charging infrastructure layout and planning for plug-in electric vehicles. In: *Cable Based and Wireless Charging Systems for Electric Vehicles* (ed. P. Sanjeevikumar), 1–24. IET.

5 Christen, D., Tschannen, S., and Biela, J. (2012). Highly efficient and compact DC-DC converter for ultra-fast charging of electric vehicles. In: *15th International Power Electronics and Motion Control Conference, EPE/PEMC 2012 ECCE Europe, Novi Sad, Serbia*.

6 Koornneef, J., Nierop, S., Saehr, H., and Wigand, F. (2015). International Comparison of Fossil Fuel Efficiency and CO_2 Intensity – Update 2015 Final Report. Ecofys.

7 Bai, S. and Lukic, S.M. (2013). Unified active filter and energy storage system for an MW electric vehicle charging station. *IEEE Transactions on Power Electronics* 12: 5793–5803.

8 IEC 62893-3:2017 (2017). *Charging cables for electric vehicles for rated voltages up to and including 0.6/1 kV – Part 3: Cables for AC charging according to modes 1, 2 and 3 of IEC 61851-1 of rated voltages up to and including 450/750 V.*

9 IEC 62893-1:2020 (2020). *Charging cables for electric vehicles for rated voltages up to and including 0.6/1 kV – Part 1: General requirements.*

10 IEC 62893-4-1:2020 (2020). *Charging cables for electric vehicles for rated voltages up to and including 0.6/1 kV – Part 4-1: Cables for DC charging according to modes 4 of IEC 61851-1 – DC charging without use of a thermal management system.*

11 IEC 62893-4-2:2021 (2021). *Charging cables for electric vehicles for rated voltages up to and including 0.6/1 kV – Part 4-2: Cables for DC charging according to modes 4 of IEC 61851-1 – Cables intended to be used with a thermal management system.*

12 IEC 62196-1:2014 (2014). *Plugs, socket-outlets, vehicle connectors and vehicle inlets – Conductive charging of electric vehicles – Part 1: General requirements.* 12-15 IEC Standards.

13 IEC 62196-3:2014 (2014). *Plugs, socket-outlets, vehicle connectors and vehicle inlets – Conductive charging of electric vehicles – Part 3: Dimensional compatibility and interchangeability requirements for d.c. and a.c./d.c. pin and contact-tube vehicle couplers.* 12-15 IEC Standards.

14 SAE J1772:2016 (2016). *SAE electric vehicle and plug-in hybrid electric vehicle conductive charge coupler*. SAE Standards.

15 Ronanki, D. and Williamson, S.S. (2018). Evolution of power converter topologies and technical considerations of power electronic transformer-based rolling stock architectures. *IEEE Transactions on Transportation Electrification* 4: 211–219.

16 https://www.aupins.com/products/ccs-combo-1-sae-j1772-dc-charging-connector/ (accessed 7 February 2023).

17 IEC 61851-24:2014 (2014). *Electric vehicle conductive charging system – Part 24: Digital communication between a d.c. EV charging station and an electric vehicle for control of d.c. charging*. 12-15 IEC Standards.

18 http://senkumachinery.com/product/-chademo-plug.html (accessed 7 February 2023).

19 IEEE Std 2030.1.1-2021 (2021). *IEEE Standard for technical specifications of a DC quick and bidirectional charger for use with electric vehicles*. IEEE Standards.

Annexure 1 Photos

Figure A1.1 shows the GB/T standard vehicle inlet and coupler. The left one is the vehicle inlet and the right one is the coupler connected to the vehicle inlet.

Figure A1.2 shows the status of the charging process of the charger.

Figure A1.3 shows the installation of a DC fast charger in an outdoor environment. The charger has two charging guns and can able to charge two EVs parallelly.

Figure A1.4 shows the charging gun connected to the electric bus.

Figure A1.5 shows the name plate rating of 120 kW DC charger.

Figure A1.6 shows the nameplate rating of the EV cable between DC charger and EV.

Figure A1.1 GB/T standard vehicle inlet and coupler.

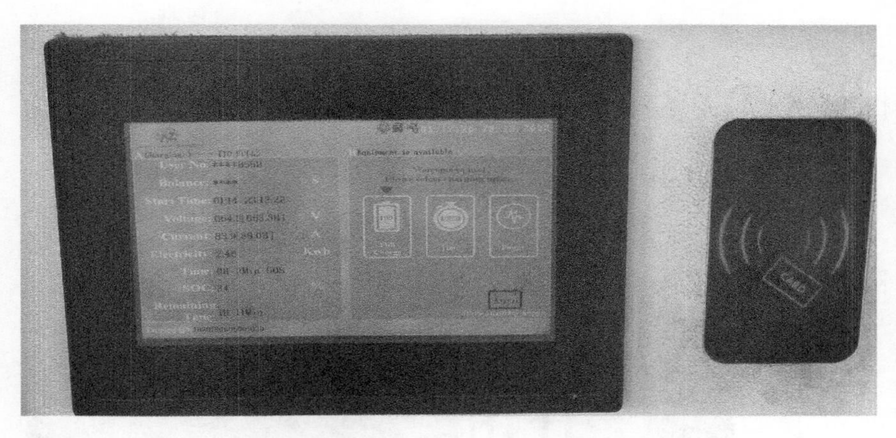

Figure A1.2 Status of the charging process of the charger.

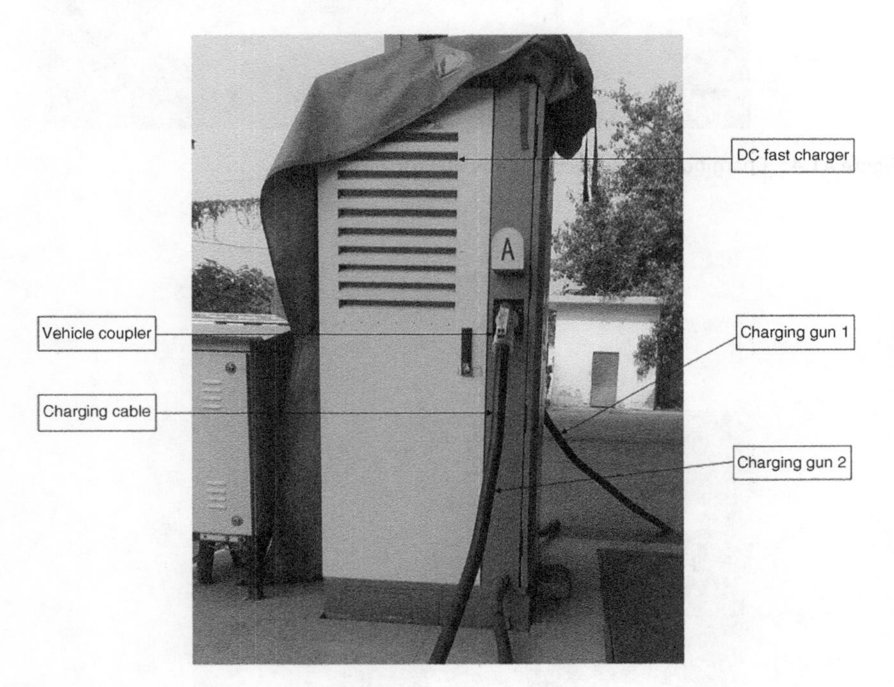

Figure A1.3 Installation of DC fast charger in outdoor.

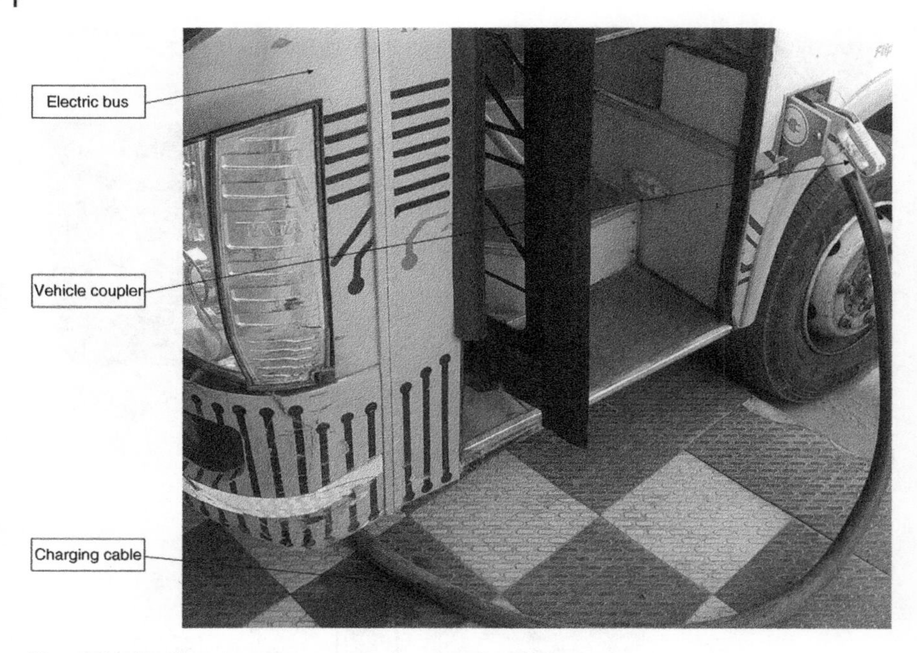

Figure A1.4 Charging gun connected to the electric bus.

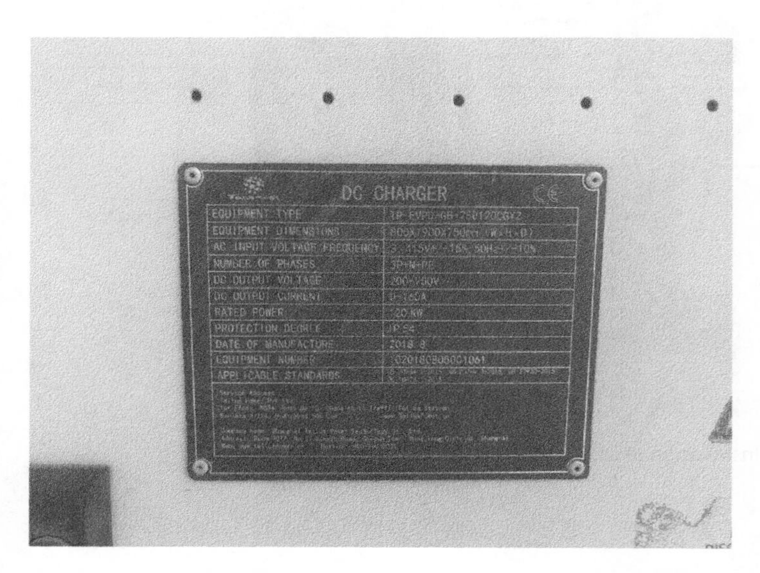

Figure A1.5 Name plate rating of DC charger. *Source:* Courtesy: Tellus Power.

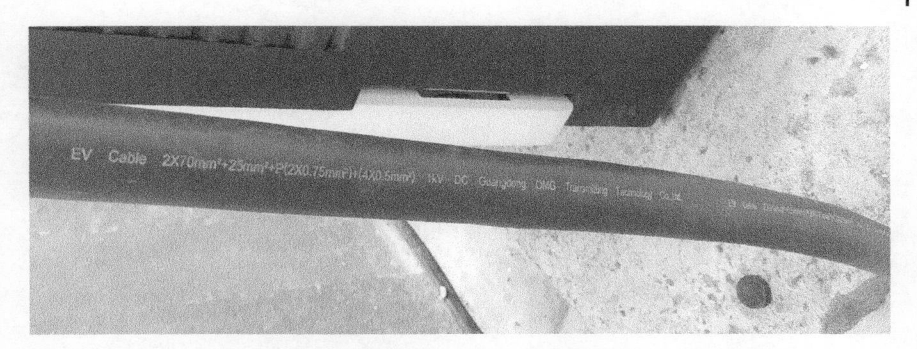

Figure A1.6 Name plate rating of EV cable. *Source:* Courtesy: Guandong OMG Transmitting Technology Co., Ltd.

2

Selection of Fast-Charging Station

2.1 Introduction

An electric vehicle (EV) becomes more popular because its operation does not need petroleum-based fuel for its operation and also it will not produce any carbon emissions. Hence, EVs are generally called as CLEANER transportation and gets more attention nowadays than any mode of transportation across the world. Many governments and EV manufacturers are taking good efforts to encourage the EV users for transportation. One of the important concerns associated with EV transportation is recharging, i.e. EV recharging may take several minutes to hours and EV users may not feel happy/comfortable about the time taken for recharging. This inconvenience (i.e. more time required for recharging the EV) is one of the major barriers that users think of before purchasing EVs and it affects the EV promotion into the market. In order to increase the number of EVs into the market, it is important to install the EV fast-charging infrastructure (i.e. charging station) in strategic locations across the country. Fast-charging stations use the high-power rating DC fast chargers to recharge the EVs in lesser time duration. The typical power rating of fast chargers is 50–150 kW for light motor vehicles and more than 150 kW for electric buses and heavy motor vehicles.

Charging the EVs with a high-power rating DC fast charger reduces the waiting time at fast-charging stations for recharging the EV batteries. The EVs users will be comfortable with recharging their vehicle in a fast-charging station whenever they need to recharge, and this will increase the number of EVs into the market. Hence, proper planning and development of fast-charging stations are required to increase the number of EVs into the market.

Fast-Charging Infrastructure for Electric and Hybrid Electric Vehicles: Methods for Large-Scale Penetration into Electric Distribution Networks, First Edition. Sivaraman Palanisamy, Sharmeela Chenniappan, and Sanjeevikumar Padmanaban.

Published 2023 by John Wiley & Sons, Inc.

In terms of users' point of view, EV charging is classified into the following two types:

- Charging at destination point (destination charging)
- Urgent charging or top-up charging

The destination charging means, the recharging happens when an EV reaches its destination point like home (using home charging), office (using workplace charging), etc. In destination charging, most of the time EV users are not worried about the recharging duration. Because they have already reached their destination and they will again use the EVs after some time (typically a few hours later). Hence, slow chargers are widely used for recharging the vehicle in destination charging.

Urgent charging or top-up is required whenever the EVs are still on the road (i.e. still EVs have to travel a few or more kilometers to reach their destination) and its battery state of charge (SoC) is reduced below the threshold value or not sufficient to reach their destination. In order to reach their destination, their EVs need to be charged essentially in between. Public charging stations are developed to recharge EVs in this regard. Users are expecting to recharge their EVs in a lesser time duration whenever they are on the road and fast chargers in charging stations are used for this purpose. In order to increase the large number of EVs into the market, a large number of public fast-charging stations need to be installed and made available for the user at any time. Hence, proper planning and development of fast-charging stations at the strategic location are essential for large EV penetration into the market.

The business model plays an important role in the planning, development, installation, and operation of any fast-charging stations. This chapter discusses the selection of the business model, the location of a fast-charging station, and electric power supply arrangements.

2.2 Business Model for Fast-Charging Stations

Whenever we think about the development of EV charging stations, the first question that comes to our mind is "who will take care of planning and implementation of public fast-charging station?" and the second question is "who will own it?"

Answers for the above two questions are varying from country to country and the business model they have been selected or followed.

Now the question comes to everyone's mind is "what is the business model for a fast-charging station?"

The business model defines **"how the EVSE services are deployed in fast-charging stations and how it is offered to their customers or users."**

The business model for fast-charging stations shall have the answers to the following questions:

1) Who owns the EVSE?
2) Who installs the EVSE?
3) Who manages or operates the EVSE?
4) What is the role of government?
5) What is the role of DISCOMs?
6) What is the role of the private sector?
7) How the EVSE is financed?
8) How do the customers pay for charging?
9) How is money earned from EVSE?

There are two different types of business models, namely the integrated model and the independent model that are widely used across the world. The business model has to be selected for the particular site or individual basis.

The business model adopted for a fast-charging station is more important because it directly relies on the revenue back to the developer/owner. Most of the countries are following any of or a mixture of the two-business listed below:

- Integrated model
- Independent model

In the integrated business model, electric utility companies (i.e. DISCOMs) will plan, install, and operate the charging stations either directly by them or partially through their contractors. In this model, all assets of the charging station are owned by DISCOMs. The advantage of using this model is that DISCOMs do not need to worry about low business volume in the starting time (initial years) because the assets are made under the regulated CAPEX of DISCOMs. The countries like Ireland and Canada (British Columbia) are using this business model.

In the independent business model, private/public–private partnership parties or companies will set up the charging stations. They will obtain the license from municipalities or local governments, and they directly manage to charge the station by themselves or may appoint the service provider to operate the charging station and payment clearance, etc. The countries like Netherlands and the United Kingdom are using this business model.

Refer to Chapter 3 for more details on the business model of fast-charging stations.

2.3 Location of Fast-Charging Station

Generally, the location of the public fast-charging stations is selected based on several factors like the number of EVs or fleets operating in that area, availability of land for establishing the fast-charging station, availability of electric power

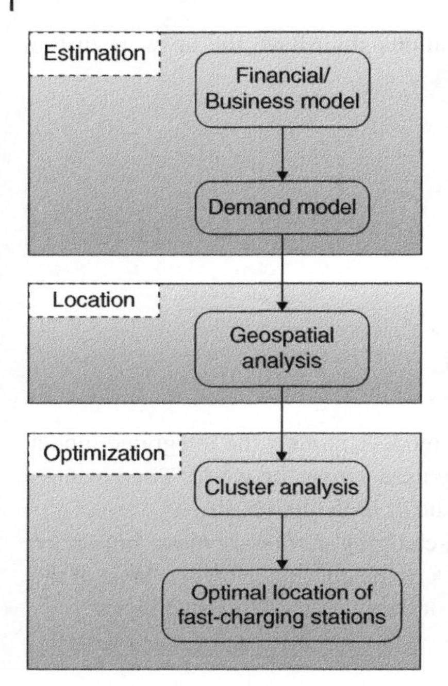

Figure 2.1 Overview of methodology.

supply, availability of network service provider, accessibility to the charging stations, etc. The typical location of the fast-charging station are highways on road, public gathering places like shopping malls, hotels, offices, hospitals, metro, and railway stations, cinema theaters, educational institutions, temples, religious places, etc.

The good planning approach for the selection of the location of fast-charging stations includes the business model that is adopted, geospatial analysis for identification of location, the total number of charger/charging station requirements within the planning area, grid impact assessments such as power and energy demand requirements, distribution system infrastructure requirements [1, 2].

The general methodology used for identifying the location of charging stations [3] is shown in Figure 2.1.

Figure 2.1 shows the three main steps (estimation, location, and optimization) involved in identifying the fast-charging stations in the planning area.

The business model adopted and demand models (types of EVs and their quantity) are the essential inputs required for the number of charging stations and their locations in the planning area. Geospatial analysis is used for identifying the locations for fast charging based on Geographical Information System (GIS) mapping and the various spatial parameters.

The step-by-step procedures for conducting the geospatial analysis are:

Step 1: Grids formation

Divide the total planning area into equal sizes of the square grid.

There are no defined requirements for the size of the grid. The best practice would be a 1×1 kM to 3×3 kM square size grid.

Step 2: Spatial parameters

Identification of various spatial parameters and assigning the values to the grid cells. The typical spatial parameters are:

- Population
- Fuel stations or gasoline stations

- Public transit stations
- Local businesses
- Office/Workplaces/Industries
- Institutions
- Point of interest/public gathering locations
- Income rates
- Park areas or parking locations
- Land values

Step 3: Normalize the grid cells
Normalize all the grid cells in the range between 0 and 1.

Step 4: Weightage
Assign the weights based on their impact on potential charging demand to different spatial parameters.

Step 5: Demand cluster
Calculate the demand clusters for each grid.

Step 6: Identification of sites
Identification of accessible sites for the installation of a fast-charging station.

This geospatial analysis can be performed through software like QGIS, ArcGIS, etc. The output of geospatial analysis provides the requirements of a total number of fast-charging stations and their locations in the planning area.

2.4 Electric Supply for Fast Charging

The developer or operator or service provider of a fast-charging station can arrange the electricity for recharging EVs in multiple ways. Based on the particular site feasibility, the fast-charging station developer or operator or service provider can select the best suitable option or combination of multiple options to meet their electricity requirements [4]. The first step in arranging the electric power supply to any fast-charging station is estimation of charging station power and energy demand requirements. This demand is the sum of all the loads (including EV charging loads, auxiliary loads, and battery swapping loads if any) of the planned fast-charging station of a particular location. Once the required power and energy demand of a fast-charging station is known, the fast-charging station developer or operator or service provider may select one or more following options to meet their requirements:

- Drawing electricity from an existing power infrastructure or connection from DISCOM.
- Getting the electricity from a new service connection from DISCOM.

- Install the small captive plant or renewable energy generation system locally on-site or closer to the fast-charging station.
- Purchase the energy from the open access market at a competitive price.

The statutory requirements of a particular country and/or state have to be followed in this regard. For example, in India, some of the minimum requirements of a public fast-charging infrastructure are as follows [5]:

- Dedicated distribution transformer with all necessary equipment and accessories.
- 33/11 kV cables or lines.
- At least two chargers minimum 100 kW (with 200–100 V), one with CHAdeMO and other with CCS specification.
- Installation of power quality meters at point of common coupling (PCC).

The DISCOMs are responsible for providing electricity to any fast-charging station within their coverage/distribution limit/boundary. Every DISCOM shall have the policies and framework, supply rules and regulations, and tariff structure for the fast-charging station.

2.5 Availability of Land

The availability of land and its accessibility is one of the important factors for the installation of a fast-charging station. Ideal places for fast-charging stations are on the main roads, main junction areas on the road, public gathering places, etc. Because, EV users prefer to charge their vehicles on the way to their destination, at their stopping places like shopping malls, movie theaters, etc. It is inconvenient for EV users if the fast-charging stations are not easily accessible. If the fast-charging stations are a little away from the main road or accessibility to the station is poor, then EV users do not prefer those charging stations [6].

Hence, it is very important to select the land/location for establishing the fast-charging station. Generally, the cost of the land is high in the metro cities. This will increase the initial capital cost of fast-charging stations.

On the other hand, the cost of land is comparatively less in non-metro cities, on the outskirts of metro cities, and on the highways.

Generally, people frequently or occasionally visit the public gathering places like shopping malls, movie theaters, temples, etc. Most of these public gathering places have parking for vehicles. In these places, the existing/available land for parking can be used for EV charging and additional land may not be required in many of these places [7–9]. Similarly, commercial establishments, office premises, and industrial plants have parking for their employee's vehicles. Also, many other vehicles (including freight vehicles) visit these places frequently for the transfer of goods, employee pickups/drops, etc. In these places, the existing parking space can be used for EV charging and additional land may/may not be required. In this regard, government policies are playing an important role in establishing fast-charging infrastructures at public gathering places, commercial establishments, and industries.

The policies of the country and/or state provide the requirements for charging infrastructures. For example, the Government of Delhi in India directed commercial and institutional buildings where the parking capacity is more than 100 vehicles have to provide 5% of their parking for charging of EVs. Another example, guidelines provided by the Ministry of Housing and Urban Affairs in India state that 20% of parking spaces shall be provided to EV charging infrastructure in all new building bylaws.

2.6 Conclusion

This chapter discussed the selection of fast-charging stations' business models, location of installation, geospatial analysis, electric power supply to fast-charging stations, and availability of load. There are two different types of business models, namely the integrated model and the independent model that are widely used across the world. The business model has to be selected for the particular site or individual basis. Geospatial analysis is used to identify the total number of charging stations and their location in the planning area. This analysis uses various spatial parameters like population, availability of electricity, etc., for the analysis. The electric power supply to fast-charging stations can be purchased in an open access competitive market or using renewable energy sources locally at the site location.

References

1 Adenaw, L. and Liekamp, M. (2020). A model for the data-based analysis and design of urban public charging infrastructure. In: *15th International Conference on Ecological Vehicles and Renewable Energies (EVER)*, 1–14.

2 Yang, W., Liu, W., Chung, C.Y., and Wen, F. (2021). Joint planning of EV fast charging stations and power distribution systems with balanced traffic flow assignment. *IEEE Transactions on Industrial Informatics* 17 (3): 1795–1809.

3 Khalife, A., Fay, T.-A., and Gohlich, D. (2022). Optimizing public charging: an integrated approach based on GIS and multi-criteria decision analysis. *World Electric Vehicle Journal* 12 (131): 1–27.

4 Zhang, H., Zechun, H., Zhiwei, X., and Song, Y. (2016). An integrated planning framework for different types of PEV charging facilities in urban area. *IEEE Transactions on Smart Grid* 7 (5): 2273–2284.

5 Charging Infrastructure for Electric Vehicles – Guidelines and Standards, Ministry of Power, pp. 1–7. New Delhi, India, 14 January 2022. https://powermin.gov.in/sites/default/files/Final_Consolidated_EVCI_Guidelines_January_2022_with_ANNEXURES.pdf (accessed 12 March 2023).

6 Asna, M., Shareef, H., Achikkulath, P. et al. (2021). Analysis of an optimal planning model for electric vehicle fast-charging station in Al Ain City, United Arab Emirates. *IEEE Access* 9: 73678–73694.

7 Ting, W., Wei, X., Zhang, X. et al. (2022). Carbon-oriented expansion planning of integrated electricity-natural gas systems with EV fast-charging stations. *IEEE Transactions on Transportation Electrification* 8 (2): 2797–2809.

8 Zhang, W., Zhao, H., and Song, Z. (2021). Integrated transit route network design and fast charging station planning for battery electric buses. *IEEE Access* 9: 51604–51617.

9 Sivaraman, P., Sharmeela, C., and Logeshkumar, S. (2021). Charging infrastructure layout and planning for plug-in electric vehicles. In: *Cable Based and Wireless Charging Systems for Electric Vehicles* (ed. P. Sanjeevikumar), 1–24. IET.

3
Business Model and Tariff Structure for Fast-Charging Station

3.1 Introduction

An electric vehicle (EV) has become more popular because its operation does not need petroleum-based fuel for its operation and it will not produce any carbon emissions [1, 2]. Hence, EVs are generally called CLEANER transportation and get more attention nowadays than any mode of transportation across the world. The governments and EV manufacturers are taking good efforts to encourage EV development for transportation. One important concern associated with EV transportation is recharging, i.e. EV recharging may take several minutes to hours and EV users may not feel happy/comfort about the time taken for recharging. This inconvenience (i.e. more time required for recharging the EV) is one of the major barriers that users think before purchasing EVs and affecting the EV promotion into the market. In order to increase the number of EVs into the market, it is important to install the EV fast-charging infrastructure (i.e. charging station) in strategic and easily accessible locations across the country. The users will comfortable with recharging their vehicles in a charging station whenever they need it, and this will increase the number of EVs into the market. Hence, proper planning and development of fast-charging stations are required to increase the number of EVs into the market.

In terms of users' point of view, EV charging is classified into the following two types:

- Charging at destination point (destination charging)
- Urgent charging

Destination charging means the recharging happens when an EV reaches its destination point like home (home charging), office (workplace charging), etc. In destination charging, most of the time EV users are not worried about the

Fast-Charging Infrastructure for Electric and Hybrid Electric Vehicles: Methods for Large-Scale Penetration into Electric Distribution Networks, First Edition. Sivaraman Palanisamy, Sharmeela Chenniappan, and Sanjeevikumar Padmanaban.

Published 2023 by John Wiley & Sons, Inc.

recharging duration. Because they have already reached their destination and they will again use the EV after some time (typically a few hours later). Hence, slow chargers are widely used for recharging the vehicle in destination charging.

Urgent charging is required whenever the EVs are still on the road (i.e. still EVs have to travel few or more kilometers to reach their destination) and its battery state of charge (SoC) is reduced below the threshold value. In order to reach their destination, their EVs essentially need to be charged. Public charging stations are developed to recharge EVs in this regard. Users are expected to recharge their EVs in a lesser time duration whenever they are on the road and fast chargers in charging stations are used for this purpose. In order to increase a large number of EVs in the market, a large number of public fast-charging stations need to be installed and made available for the user at any time.

Fast-charging stations are used to charge the EVs in a lesser time duration. These stations use the electricity from DISCOMs and/or on-site renewable power that is used to charge the batteries. Based on the amount of energy transfer (i.e. kWh) from the charging stations to EV batteries, these fast-charging stations charge the EV users, i.e. EV users will pay for the actual energy consumed by the battery. Fast-charging station developers or operators are making revenue by charging EVs. The business model adopted for the fast-charging station plays an important role in the charging station infrastructure business. Because the revenue coming from the fast-charging station depends on the business model adopted.

Now the question comes to everyone's mind is "what is the business model for fast-charging stations?"

The business model defines how the EVSE services are deployed in fast-charging stations and how it is offered to their customers or users [2].

The business model for fast-charging stations shall have the answers to the following questions:

1) Who owns the EVSE?
2) Who installs the EVSE?
3) Who manages or operates the EVSE?
4) What is the role of government?
5) What is the role of DISCOMs?
6) What is the role of private sectors?
7) How the EVSE is financed?
8) How do the customers pay for charging?
9) How is money earned from EVSE?

There are two different types of business models, namely the integrated model and the independent model that are widely used across the world [3]. The business model has to be selected for the particular site or individual basis.

3.2 Business Model

The rapid or mass adoption of EVs can happen when the business model for both EVs and their charging infrastructure is cost-effective and beneficial to the parties involved in the EV ecosystem. The availability of public fast-charging stations is the expectation or requirement of EV users and that has to be available across the country. The governments in various countries providing grants and tax benefits for transition to EVs could help the EV industry to take-off and boost the early adapters or users of EVs for various applications [4]. The typical EV fast-charging infrastructure layout is shown in Figure 3.1.

The business model adopted for a fast-charging station is more important because it directly relies on the revenue back to the developer/owner. Most of the countries are following any of or a mixture of the two businesses listed below [3]:

- Integrated model
- Independent model

3.2.1 Integrated Model

In the integrated business model, electric utility companies (i.e. DISCOMs) will plan, install, and operate the charging stations either directly by themselves or through their contractors. In this model, all assets of the charging station are owned by DISCOMs [3]. Figure 3.2 shows the typical layout of the fast-charging station using the integrated business model directly owned by DISCOMs. In this model, planning, installation, and operation of the fast-charging stations are directly performed by DISCOMs [4].

Figure 3.3 shows the typical layout of the fast-charging station using the integrated business model directly owned by DISCOMs and operated by their contractors. In this model, planning and installation are directly performed by DISCOMs, and operation of the fast-charging stations is performed by contractors of DISCOMs [4].

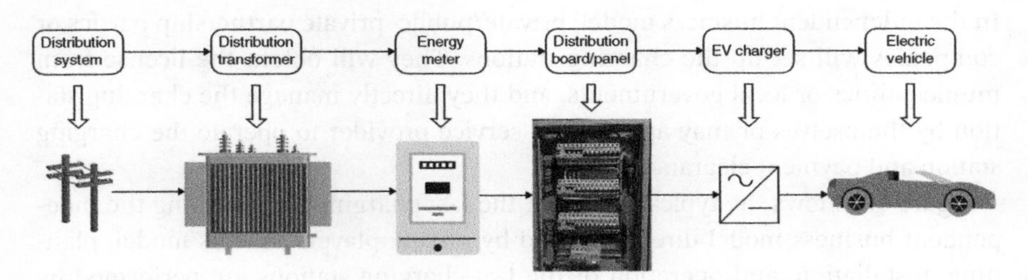

Figure 3.1 Typical EV fast-charging infrastructure layout.

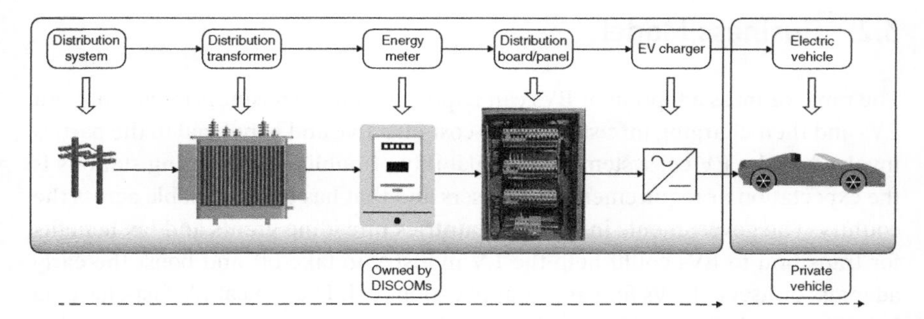

Figure 3.2 Typical layout of fast-charging station using the integrated business model directly owned by DISCOMs.

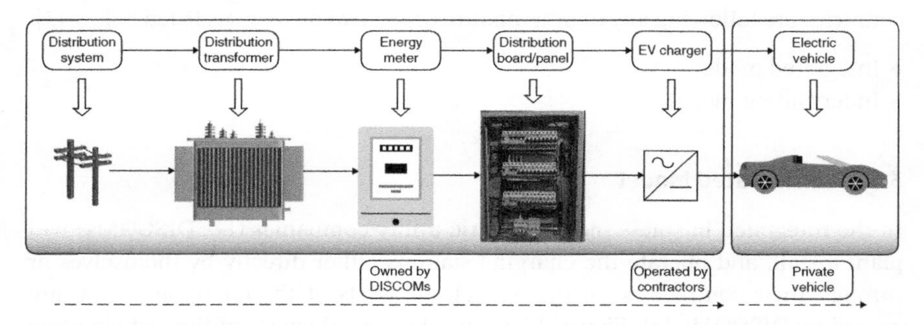

Figure 3.3 Typical layout of fast-charging station using the integrated business model directly owned by DISCOMs and operated by their contractors.

The advantage of using this model is that DISCOMs do not need to worry about low business volume in the starting time (initial years) because the assets are made under the regulated capital expenditure (CAPEX) of DISCOMs. The countries like Ireland and Canada (British Columbia) are using this business model [3].

3.2.2 Independent Model

In the independent business model, private/public–private partnership parties or companies will set up the charging stations. They will obtain the license from municipalities or local governments, and they directly manage the charging station by themselves or may appoint the service provider to operate the charging station and payment clearance, etc. [3].

Figure 3.4 shows the typical layout of the fast-charging station using the independent business model directly owned by private players. In this model, planning, installation, and operation of the fast-charging stations are performed by private players [4].

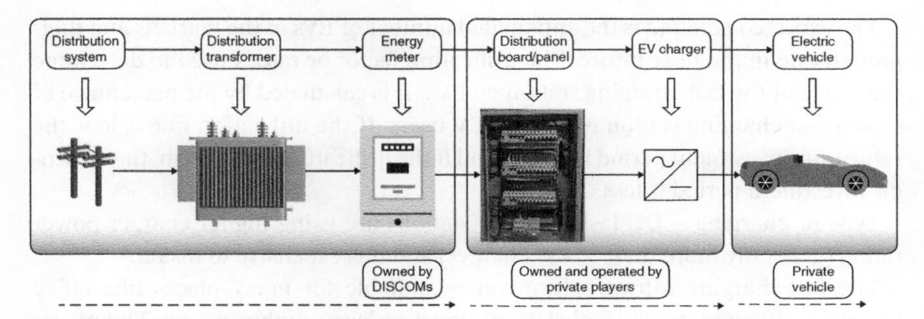

Figure 3.4 Typical layout of the fast-charging station using the independent business model owned by private players.

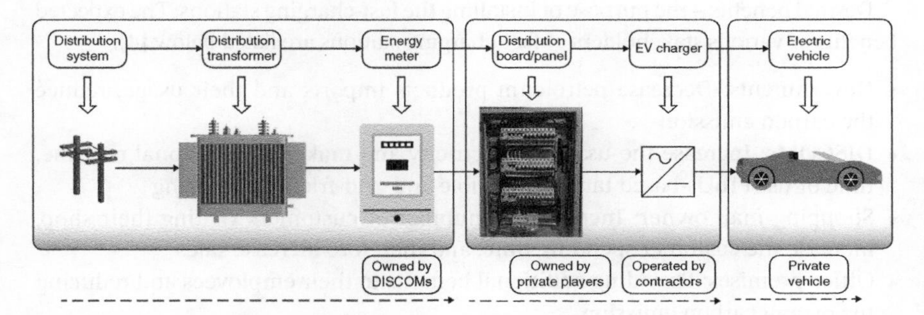

Figure 3.5 Typical layout of the fast-charging station using the independent business model owned by private players and operated by their contractors.

Figure 3.5 shows the typical layout of the fast-charging station using the independent business model directly owned by private players and operated by their contractors. In this model, planning and installation are carried out by private players, and the operation of the fast-charging stations is performed by contractors of private players [4].

The countries like Netherlands and the United Kingdom are using this business model.

3.2.3 Selection of Business Model for Fast-Charging Station

The following factors are considered while selecting the business model for fast-charging stations [4]:

- Expected demand
- Type of charging
- Location
- Desired benefits

The expected demand is the anticipated number of EVs in the markets and their growth rate in the near future. The utilization factor or rate is used to determine the usage of the fast-charging station by EVs. It is calculated by the percentage of time a fast-charging station is used by EV users. If the utilization rate is low, the return-on-investment period is longer and if the utilization rate is high, the return-on-investment period is less.

Type of charging – DC fast-charging stations are using higher charger power rating (generally more than 50 kW charger) and it is expensive to install.

The fast-charging infrastructure can be suitable for many places like office premises, shopping malls, fuel stations, tourist places, highways, etc. The power rating of the charger for these locations is based on how fast the charging process needs to complete.

Desired benefits – the purpose of installing the fast-charging stations. The expected benefits of various stakeholders of fast-charging stations are listed below [4]:

- Governments: Decrease petroleum products imports and their usage, reduce the carbon emission
- DISCOMs: Increase the usage of electricity and make the additional revenue, time of use (ToU)-based tariff for flexible and grid-friendly charging
- Shopping mall owner: Increase the number of customers visiting their shop, increase the customer spending time, and therefore increase sales
- Office premises: Providing additional benefits to their employees and reducing the overall carbon emission

3.2.4 Fast-Charging Infrastructure and Operating Expenses

Any fast-charging station shall have the necessary infrastructure to perform EV charging. The expenses for establishing the fast-charging stations and their operations are classified into capital expenses and operational expenses [4–6].

A) Typical fast-charging stations capital investments are:
   - Administrative expenses
   - Grid infrastructure outside of fast-charging station expenses
   - EVSE within the fast-charging station expenses
   - Construction expenses

   The typical administrative expenses are:
   - Land identification or searching
   - Land acquisition or land purchase cost
   - Obtaining the necessary permissions from concerned authorities
   - Financing

   The typical grid infrastructure outside of fast-charging station expenses are:
   - Distribution transformer

- Substation equipment
- Cables and conductors
- Energy meters and power quality meters
- Other: grid connection expenses

The typical EVSE within the fast-charging station expenses are:

- Electrical distribution boards or panels
- Cables
- Circuit breakers
- Charger units
- On-site renewable energy generation (if applicable)
- Signage boards and safety equipment

The typical construction expenses of the fast-charging station are:

- Laborers for construction
- Temporary equipment for construction
- Temporary power supply for construction
- Equipment's transportation costs

B) Typical fast-charging stations operational expenses are [6]:
- Cost for electricity including energy consumption and purchased demand cost
- Owner payment
- Site maintenance cost including labor cost, maintenance costs like snow removal, insurance, etc.

3.3 Battery Swapping

Battery swapping is a method to replace the EV's drained battery with a fully charged battery. Battery replacement can be done in a few minutes either manually or automatically by using robots like refilling the fuel in fuel stations for conventional vehicles. This method is saving the waiting time required for recharging the vehicle at charging stations. The battery swapping can be done at battery-swapping stations and/or fast-charging stations with the facilities for battery swapping. Once the drained battery is replaced by a fully charged battery, the drained battery will be housed on the battery swapping station charging port. Once these drained batteries at battery swapping stations are fully charged, then the next EVs user can use it.

There are some challenges involved in battery swapping method, they are:

1) Standardization of battery characteristics such as
   - Voltage (V) of the battery
     The voltage rating of both batteries (drained battery in the EV and fully charged battery available at battery swapping station) should be the same, so that battery swapping can be done.

- Capacity (kWh) of the battery pack
  The battery capacity in kWh of both batteries (drained battery in the EV and fully charged battery available at battery swapping station) should be the same, so that battery swapping can be done. In case of any difference in capacity, it would be difficult to swap the battery. If the EV battery rating is higher than the battery available at the battery swapping station, the EV user may not prefer to swap the battery.
- C rating
  C rating of the battery is used to describe the charging and discharging characteristics of the battery. C rating of both batteries is to be the same.
- Number of cells or modules in a battery pack
  A battery pack used in EV consists of multiple cells or modules at achieving the rated voltage and capacity. The number of cells or modules of both batteries is to be the same.

2) Rating of battery capacity is varied from manufacturer to manufacturer
   - In order to attract users, every EV manufacturer is offering a different travel range. So, the battery capacity and its characteristics are varied from manufacturer to manufacturer. Hence, the requirements of battery swapping stations are increased to cater to all EV manufacturer's vehicles.
3) Battery technology is varied from manufacturer to manufacturer
   - The battery technology (Li-ion, lead acid, NMH, etc.) used in the EVs also varied from manufacturer to manufacturer.
4) Cost of the battery is varied from manufacturer to manufacturer
   - The cost of the battery is varied from manufacturer to manufacturer for the same battery technology and same rating.
5) Lifetime of the battery
   - The batteries are providing better or good performance for the specified period of time. After that time period, battery performance gets affected and EV users have to change the battery with a newer one. The EV users may change the battery, which has completed its lifetime, with the newer battery available in the battery swapping stations. Similarly, the lifetime-completed battery in the battery swapping stations may be replaced with a new battery for EVs. In both cases, either EV users or battery-swapping station operators get affected.

The business model for the battery swapping method has to address the challenges involved in adopting this model on a large scale. Policies play an important role in this. Every country shall have a policy in place that specifies the battery technology (i.e. battery chemistry), battery pack capacity (kWh), etc., to be used in the country. The standardization of various battery parameters can motivate EV users to use the battery swapping method whenever they are in travel to reduce the weighting time required at charging stations.

3.4 Tariff Structure

The electricity tariff structure specifies how consumers are charged for using electricity [7]. Generally, the fast-charging stations have two different tariffs. They are:

1) Tariff between electric utilities (DISCOMs) and fast-charging stations.
2) Tariff between fast-charging stations and EV users.

3.4.1 Tariff Between Electric Utilities (DISCOMs) and Fast-Charging Stations

The DISCOMs are collecting revenue from the fast-charging stations for using the electricity. The tariff structure varies from DISCOMs to DISCOMs for fast-charging stations. The typical tariff structure of DISCOMs is listed below and DISCOMs may collect one or more charges [7].

- Energy charges: The DISCOMs collect the revenue based on the energy (in kWh) consumed by fast-charging stations. Many DISCOMs have a different rate for energy consumption ($/kWh) based on the ToU.
- Demand chargers: The DISCOMs charge the fast-charging stations based on the maximum demand or contract demand in $/kVA or $/kW.
- Connection chargers: The DISCOMs collect a fixed rate of revenue from fast-charging stations in each billing cycle independent of the actual energy consumptions of the billing cycle.

In addition to the above charges, taxes (%) are charged by DISCOMs based on guidelines provided by the governments.

3.4.2 Tariff Between Fast-Charging Stations and EV Users

The fast-charging station operators collect revenue from EV owners based on the amount of energy (in kWh) transferred to EVs battery, i.e. energy consumed by EV ($/kWh). The EV tariff structure can be designed in such a way that, EV charging can be beneficial to both EV users as well as utility. If every EV user is charging their EVs at the same time, it increases the power demand of the DISCOMs and DISCOMs have to expand their distribution infrastructure for this. Due to this, DISCOMs may increase the cost of energy ($/kWh) supplied to fast-charging stations. ToU-based tariff structure plays an important role to disaggregate the single EV charging time into multiple times [7, 8]. Because ToU-based tariff structure provides different tariff rates ($/kWh) based on the time of usage. The tariff rate is low at the time of surplus power available in the grid and the tariff rate is high at the time of grid operating with deficient or shortage of power.

Generally, during the morning peak hours (6 a.m.–9 a.m.) and evening peak hours (6 p.m.–10 p.m.), the grid is operated with more connected loads. During this peak time period, the cost of energy is high ($/kWh). On the off-peak hours (10 a.m.–5 p.m.) in the daytime and (11 p.m. to next day 5 a.m.) in the night hours, the grid is operating with a lesser connected load as compared with loads connected in the peak hours. During this off-peak time period, the cost of energy is low ($/kWh). The EV users can select their EV charging time based on their comfort time period. Charging the EVs during off-peak hours can provide some savings to EV users.

So, the tariff structure has to encourage EV users to charge their vehicles at off-peak hours with electricity tariffs and discourage EV users to charge their vehicles at peak hours.

The tariff design objectives include the following [7, 9]:

- Ensure the cost recovery for fast-charging stations.
- Ensure the revenue stability and profitability of fast-charging stations.
- Encourage the EV users to use the ToU as energy conservation and reduce the stress on the grid infrastructure.

3.5 Conclusion

Fast-charging stations are used to charge the EVs in a lesser time duration. These stations use the electricity from DISCOMs and/or on-site renewable power that is used to charge the batteries. Based on the amount of energy transfer (i.e. kWh) from the charging stations to EV batteries, these fast-charging stations charge the EV users, i.e. EV users will pay for the actual energy consumed by the battery. Fast-charging station developers or operators are making revenue by charging EVs. The business model adopted for the fast-charging station plays an important role in the charging station infrastructure business. Because the revenue coming from the fast-charging station depends on the business model adopted. This chapter discussed the business models (such as integrated business model and independent business model) and tariff structures used for EV fast-charging stations.

References

1 Sivaraman, P., Sharmeela, C., and Logeshkumar, S. (2021). Charging infrastructure layout and planning for plug-in electric vehicles. In: *Cable Based and Wireless Charging Systems for Electric Vehicles* (ed. P. Sanjeevikumar), 1–24. IET.

2 Moghaddam, Z., Ahmad, I., Habibi, D., and Phung, Q.V. (2018). Smart charging strategy for electric vehicle charging stations. *IEEE Transactions on Transportation Electrification* 4 (1): 76–88.

3 Pillai, R.K., Suri, R., Kundu, S. et al. (2018). *Electric Vehicle Charging Stations Business Models for India, Indian Smart Grid Forum*, 1–15.
4 Bopp, K., Zinaman, O., and Lee, N. (2021). Electric Vehicle Charging Infrastructure: Business Model and Tariff Design Support to the Lao PDR. Report Number NREL/PR-7A40-77671. NREL, United States. https://www.osti.gov/servlets/purl/1762451.
5 Mao, D., Tan, J., and Wang, J. (2021). Location planning of PEV fast-charging station: an integrated approach under traffic and power grid requirements. *IEEE Transactions on Intelligent Transportation Systems* 22 (1): 483–492.
6 Chittenden County RPC (2014). *Electric Vehicle Charging Station Guidebook – Planning for Installation and Operation*. Chittenden County RPC.
7 Zinaman, O., Bopp, K., Lee, N., and Beshilas, L. (2020). Electric Vehicle Supply Equipment: Tariff Design Support to the Lao PDR. Report Number NREL/PR-7A40-77747. NREL, United States.
8 GIZ Report on Integration of Electric Vehicles Charging Infrastructure with Distribution Grid: Global Review, India's Gap Analyses and Way Forward (2021). Report 2 on International Review on Integration of Electric Vehicles Charging Infrastructure with Distribution Grid.
9 Aznar, A., Belding, Scott, Bopp, K., et al. Building Blocks of Electric Vehicle Deployment: A Guide for Developing Countries. United States. https://doi.org/10.2172/1785089

4

Batteries for Fast-Charging Infrastructure

4.1 Introduction

Electric vehicles are considered an alternative to conventional vehicles and these vehicles use the energy stored in the batteries for their operation. Batteries are electrochemical devices and they are used to store energy in a chemical medium. Generally, batteries are used to perform functions, namely charging and discharging. During the charging process, the electrical energy drawn from the grid source is stored in the batteries for future use and the same energy is used to power the equipment (i.e. electric motor in the EV) during the discharging time. One of the important factors which affect the large-scale penetration of EVs into the market is the "time required for EV charging is high." Whenever the EVs are charged by an AC slow charger, it takes a few hours (typically two to six hours) for recharging the batteries from 0% SoC to 100% SoC. Many EV users are not convenient with the longer time required for charging when they are in the middle of their travel and it will delay in reaching their destination. So, it is essential to reduce the required charging time from a few hours to a few minutes. Fast charging is a method, and it is used to charge EVs in a lesser time duration as compared with AC slow chargers. Fast chargers are used for this purpose. In this fast-charging method, higher power is used to charge the batteries in a lesser time duration [1, 2].

Hence, fast charging is one of the key features for large penetration of EV adoption into the market. The batteries used in the vehicles shall be suitable for fast charging. At the same time, charging the batteries at a higher rate accelerate battery degradation like capacity degradation as well as power capability deterioration. It is important to understand that, during fast charging, batteries will generate heat due to resistive heating, and difficult to remove them in a uniform and efficient manner. Due to the nonuniformity of current distribution during the

Fast-Charging Infrastructure for Electric and Hybrid Electric Vehicles: Methods for Large-Scale Penetration into Electric Distribution Networks, First Edition. Sivaraman Palanisamy, Sharmeela Chenniappan, and Sanjeevikumar Padmanaban.

Published 2023 by John Wiley & Sons, Inc.

charging and discharging cycle, the heat generation of the battery pack is not uniform, and this leads to a temperature gradient. This leads to safety concerns for the battery and accelerates the degradation.

This chapter discusses the requirements of a battery that is suitable for fast-charging requirements.

4.2 C-Rating of the Battery

C-rating is also called Charging rating, which is important for battery charging and discharging behavior. In simple words, C-rating governs the charge and discharge rating of the battery. C-rating is a measure of the rate at which the battery can be charged and discharged relative to the battery's maximum capacity. The typical C-rating of a battery is listed in Table 4.1.

Table 4.1 Typical C-rating of a battery.

S. No	C-rating
1	5C
2	2C
3	1C
4	0.5C
5	0.2C
6	0.1C
7	0.05C

Example 4.1

If the C-rating of a battery is 1C, and the battery capacity is 1 Ah, this battery can deliver the 1 A current for one-hour duration.

Example 4.2

If the C-rating of a battery is 0.5C, and the battery capacity is 1 Ah, this battery can deliver the 0.5 A current for two-hour duration.

Example 4.3

If the C-rating of a battery is 2C, and the battery capacity is 1 Ah, this battery can deliver the 2 A current for 30 minutes duration.

In order to perform fast charging, the battery shall have a higher C-rating for charging and discharging characteristics. A higher C-rating of the battery accepts the higher current during the charging process.

4.3 Different Types of Chemistries

The typical architecture of an EV is shown in Figure 4.1.

In typical battery-operated EVs, batteries are the main source for providing the required electric power and energy to run the electric motor. These batteries are required to handle the high power required to run the motor and the high energy demand for a longer duration of travel. In battery-operated EVs, the required power and energy have to be met by the battery only, hence the size of the battery will be bigger, and it requires more space to accommodate it. The actual concern with respect to battery placement in an EV is weight and limited space. Generally, batteries are one of the bigger components in battery-operated EVs based on the component size.

An ideal battery for EVs is expected to have high power and energy densities, long life for enabling both long-range travels on a single charge, and quickly recharge whenever their battery SoC is low and not sufficient to reach their destination in any climatic/weather conditions in any location [3].

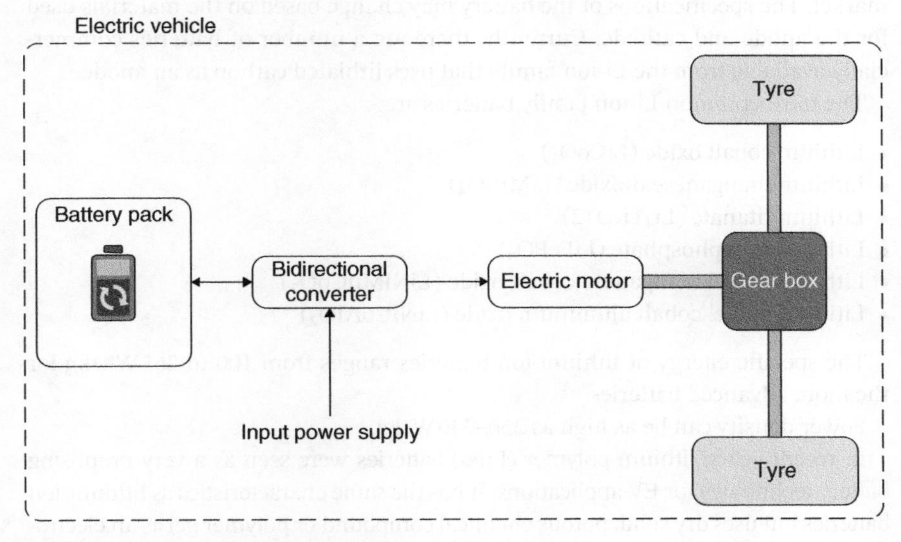

Figure 4.1 Typical architecture of EVs.

The chemistry used in the battery is important for its application. There are different types of battery chemistries available in the market and are under research and development. Some of the chemistries used in the batteries are:

- Li-ion family
- Lead acid
- Nickel family

Each battery chemistry has its own advantages, disadvantages, and limitation to use it.

4.3.1 Li-Ion Family

Lithium is the lightest metal with a molecular weight of 6.941 g/mol.

Lithium-ion (Li-ion) batteries are the widely used battery technology in plug-in EVs/HEVs and they dominate the other battery technology in terms of lesser weight and higher specific energy. Presently, most of the EVs/HEVs that are coming to the markets are using Li-ion battery packs. Li-ion batteries are considered high-performance batteries for EVs due to their lithium characteristics. As compared with other metals, lithium has a higher potential and electrochemical equivalent leading to the highest specific energy potential (Wh/kg) and on a volumetric energy basis (Wh/l). Lithium is also light in its weight and the lightest of all metallic materials [4].

The final product of the Li-ion batteries family comes in many ways into the market. The specifications of the battery may change based on the materials used for the anode and cathode. Currently, there are a number of batteries commercially available from the Li-ion family that uses lithiated carbon as an anode.

The most common Li-ion family batteries are:

- Lithium cobalt oxide ($LiCoO_2$)
- Lithium manganese dioxide ($LiMn_2O_4$)
- Lithium titanate (Li_4Ti_5O12)
- Lithium iron phosphate ($LiFePO_4$)
- Lithium nickel manganese cobalt oxide ($LiNiMnCoO_2$)
- Lithium nickel cobalt aluminum oxide ($LiNiCoAlO_2$)

The specific energy of lithium-ion batteries ranges from 100 to 265 Wh/kg for the more advanced batteries.

Power density can be as high as 250–340 W/kg.

In recent years, lithium-polymer (LiPo) batteries were seen as a very promising battery technology for EV applications. It has the same characteristics as lithium-ion batteries but uses dry solid, porous chemical compound or polymer gel as an electrolyte. The polymer battery is thinner and lighter compared with common Li-ion battery and it can be used in applications where thin-shaped batteries are required.

The comparison of general Li-ion batteries and LiPo batteries is listed in Table 4.2.

Table 4.2 Comparison of general Li-ion batteries and LiPo batteries. *Source:* Adapted from [5].

S. No	Parameter	Li-ion	LiPo
1	Electrolyte	Liquid	Solid
2	Energy density	High	Low
3	Weight	Heavier	Lighter
4	Efficiency	85–95%	75–85%
5	Aging	Fast	Slow
6	Self-discharge	Low	Very low
7	Cycle life	Shorter	Larger
8	Safety	Chances of fire, etc.	Relatively safer
9	Cost	Cheaper	Expensive

4.3.2 Lead Acid

Lead-acid (Pb/A) batteries are the oldest type of battery used in internal combustion vehicles for starting purposes. These battery-negative electrodes contain lead (Pb) and the positive plates have lead dioxide (PbO_2). Sulfuric acid (H_2SO_4) is used as an electrolyte. During the discharge time, both the positive electrode and negative electrodes react with the electrolyte (sulfuric acid). The energy is released during the chemical reaction and when energy is added, the process will reverse.

The overall reaction is:

$$Pb + PbO_2 + 2H_2SO_4 \longleftrightarrow 2PbSO_4 + 2H_2O \tag{4.1}$$

Lead acid batteries are the most mature technology for EV applications, and the costs of the batteries are the lowest. At the same time, their specific energy (Wh/kg) also is less as compared to the other battery technologies. A typical lead acid battery has a specific energy of 35 Wh/kg and a specific power of 250 W/kg. The recent valve-regulated lead acid (VRLA) batteries have 800 cycles. The charging and discharging of typical lead acid batteries and VRLA batteries are slow and are not suitable for use of fast charging of EVs/HEVs.

4.3.3 Nickel Family

The nickel-based batteries use nickel as the positive electrode of the battery and they are classified into four types, namely:

- Nickel iron (Ni-Fe)
- Nickel zinc (Ni-Zn)
- Nickel cadmium (Ni-Ca)
- Nickel metal hydride (Ni-MH)

The Ni-Fe and Ni-Zn batteries are not a suitable option for EV application due to low specific power and lesser lifecycle. Ni-Ca is the mature technology of the Nickel family. The Ni-MH battery of the Nickel family is the most suitable for EV applications. The advantage of this battery is environmentally friendly because they do not use cadmium.

The overall reaction is,

$$MH + NiOOH \longleftrightarrow M + Ni(OH)_2 \tag{4.2}$$

Ni-MH batteries are used in HEVs like the Honda Civic hybrid and the Toyota Prius because of their high specific power. The specific power of Ni-MH batteries is in the range of 1200–2000 W/kg. The specific energy is in the range of 50–75 Wh/kg.

The Ni-MH batteries have a lifetime of 3000 cycles if operated between 20% and 80% SoC (state of charge) and this battery can be able to meet the projected lifetime requirements of EVs and plug-in HEVs.

Ni-MH does not deteriorate over time, like Li-ion batteries.

4.3.4 Selection of Battery Chemistry

The selection of battery chemistry for any particular EV application depends on one or more of the following factors [6]:

- Specific power for good acceleration
- Specific energy for longer-range or long-distance travel
- Peak discharge power
- Peak charge power
- Performance
- Storage and ambient temperature
- Cooling requirements
- Wider temperature operating bandwidth
- Number of charging and discharging cycle
- Safety
- Long life span
- Cost

The specific power, specific energy, energy density, and life cycle for lead acid, Li-ion, Ni-Cd, and Ni-MH batteries are listed in Table 4.3.

4.4 Batteries Used in EVs in the Market

The batteries in the EVs are the main source of energy and are used to provide the required power and energy to run the electric motor in the EVs. Table 4.4 provides the battery type and their capacity in various EVs available in the market.

Table 4.3 Specific power, specific energy, energy density, and life cycle for lead acid, Li-ion, Ni-Cd, and Ni-MH batteries. Adapted from [5, 8].

S. No	Nominal voltage (V)	Battery technology	Specific power (W/kg)	Specific energy (Wh/kg)	Energy density (Wh/l)	Life cycle
1	2.1	Lead acid	180	30–40	100	500
2	1.2	Ni-Cd	200	50–80	300	2000
3	1.2	Ni-MH	200–300	60–120	180–220	< 3000
4	3.6	Li-ion	200–430	100–265	200–600	2000

Table 4.4 Battery type and their capacity in various EVs available in the market. [7] / IEEE / Public Domain CC BY 4.0.

Manufacturer	Model	Power (kW)	Type	Battery (kWh)	Battery voltage (V)	Autonomy (km)[a]
BMW	i3	137	BEV	42.2 (Li-ion)	360	359
	I3s	137	BEV	42.2 (Li-ion)	360	344
	iX3	213	BEV	74 (Li-ion)	—[b]	360
Mercedes-Benz	EQC	304	BEV	80 (Li-ion)	405	409
	EQV	152	BEV	90 (Li-ion)	—[b]	418
Nissan	Leaf e+	162	BEV	62 (Li-ion)	384	385
	Ariya	225	BEV	87 (Li-ion)	320	500
Porsche	Taycan 4S	395	BEV	93.4 (Li-ion)	800	464
Renault	Zoe	101	BEV	52 (Li-ion)	346	390
Tesla	Model S	593	BEV	100 (Li-ion)	350	610
	Model 3	635	BEV	75 (Li-ion)	300	530
	Model X	593	BEV	100 (Li-ion)	350	487
	Model Y	465	BEV	100 (Li-ion)	350	480
Toyota	Rav4[c]	228	PHEV	18 (Li-ion)	386	65
	Prius[c]	101	PHEV	8.8 (Li-ion)	—[b]	40
Volkswagen	e-up	62	BEV	32.3 (Li-ion)	307	258
	e-Golf	101	BEV	35.8 (Li-ion)	323	198
	ID.3	152	BEV	82 (Li-ion)	323	549
	ID.4	152	BEV	77 (Li-ion)	—[b]	520
	Golf GTE	152	PHEV	13 (Li-ion)	345	40
	Passat GTE	163	PHEV	13 (Li-ion)	345	55

[a] World harmonized Light-duty vehicles Test Procedure (WLTP): harmonized standard for determining pollution, CO_2 emissions, and fuel consumption of traditional and hybrid cars, as well as the range of fully electric vehicles.
[b] Data not provided by manufacturers.
[c] Models ready to go on the market in 2021.

4.5 Conclusion

An EV's batteries are used to provide the energy to run a vehicle whenever they are on the road. The energy stored in the batteries is used for this purpose. Generally, batteries are used to perform functions, namely charging and discharging. During the charging process, the electrical energy drawn from the grid source is stored in the batteries and during the discharging time, the same energy is used to run the EVs. The C-rating is an important parameter that is used to determine the charging and discharging characteristics of a battery. The fast chargers are used to charge the EV batteries in a lesser time duration. The batteries used in the vehicle shall be suitable for fast charging so that EV users can able to charge their vehicles in a lesser time duration. This chapter discussed the various battery chemistry and batteries available in the market.

References

1 Gjelaj, M., Hashemi, S., Traeholt, C., and Andersen, P.B. (2018). Grid integration of DC fast-charging stations for EVs by using modular li-ion batteries. *IET Generation, Transmission and Distribution* 12 (20): 4368–4376.

2 Sivaraman, P., Sharmeela, C., and Logeshkumar, S. (2021). Charging infrastructure layout and planning for plug-in electric vehicles. In: *Cable Based and Wireless Charging Systems for Electric Vehicles* (ed. P. Sanjeevikumar), 1–24. IET.

3 Tomaszewska, A., Chu, Z., Feng, X. et al. (2019). Lithium-ion battery fast charging: a review. *eTransportation* 1: 1–28.

4 Hannan, M.A., Hoque, M.M., Hussain, A. et al. (2018). State-of-the-art and energy management system of lithium-ion batteries in electric vehicle applications: Issues and recommendations. *IEEE Access* 6: 19362–19378.

5 Vidyanandan, K.V. (2019). Batteries for electric vehicles. *Energy Scan: A House e-Journal of Corporate Planning*, NTPC Ltd. I (38): 1–7, New Delhi.

6 Sivaraman, P. and Sharmeela, C. (2020). IoT-based battery management system for hybrid electric vehicle. In: *Artificial Intelligent Techniques for Electric and Hybrid Electric Vehicles* (ed. A. Chitra, P. Sanjeevikumar, J.B. Holm-Nielsen, and S. Himavathi), 1–16. Scrivener Publishing LLC.

7 Aretxabaleta, I., De Alegria, I.M., Andreu, J. et al. (2021). High-voltage stations for electric vehicle fast-charging: trends, standards, charging modes and comparison of unity power-factor rectifiers. *IEEE Access* 9: 102177–102194.

8 https://en.wikipedia.org/wiki/Lithium-ion_battery (accessed in 06-05-2023).

5

Distribution System Planning

5.1 Introduction

The distribution system is the last stage of the power system, and it is powering the large number of end users and their equipment (loads). The loads connected in the distribution systems are generally classified into residential, commercial, and industrial system loads [1]. The general structure of the distribution system is shown in Figure 5.1. The distribution systems are classified into two types, namely primary distribution and secondary distribution. It consists of primary distribution lines, secondary distribution lines, distribution transformers, capacitor banks and harmonic filters, medium voltage and low voltage switchgears, etc. [2, 3]. Medium and large industrial customers are directly powered by medium voltage through primary distribution feeders. The distribution transformers are used to step down the medium voltage level to the low voltage level. The end users such as residential, commercial, and small-scale industrial user loads are connected at the low voltage side.

The operating loads of residential, commercial, and small-scale industrial systems are not uniform or constant and vary throughout their operation in a day. The typical load curve of a distribution system is shown in Figure 5.2.

EV charging is classified into three types, namely residential charging, workplace charging, and public charging. Personal or private vehicles (two-wheelers, three-wheelers, and four-wheelers) are primarily charged at the residents and sometimes at workplaces or public charging [4]. Most of the commercial vehicles (three-wheelers and four-wheelers) are charged at public charging stations.

Fast-Charging Infrastructure for Electric and Hybrid Electric Vehicles: Methods for Large-Scale Penetration into Electric Distribution Networks, First Edition. Sivaraman Palanisamy, Sharmeela Chenniappan, and Sanjeevikumar Padmanaban.

Published 2023 by John Wiley & Sons, Inc.

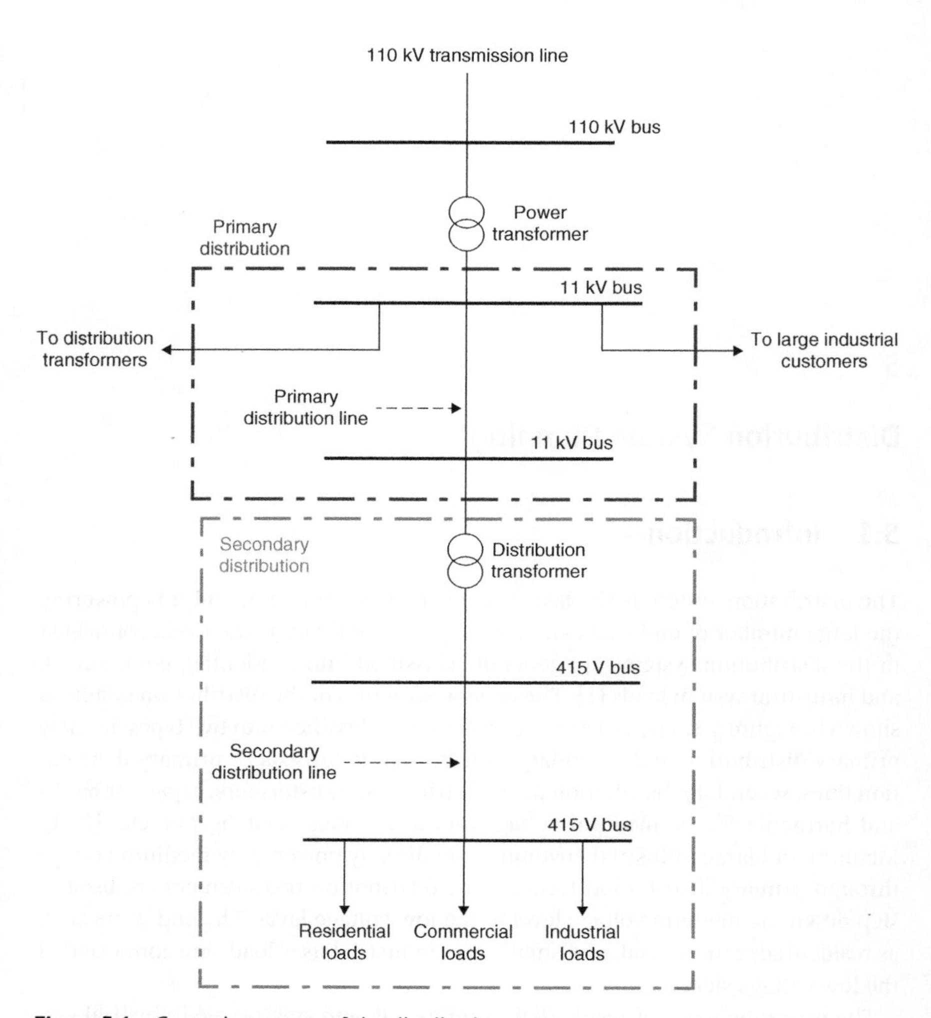

Figure 5.1 General structure of the distribution system.

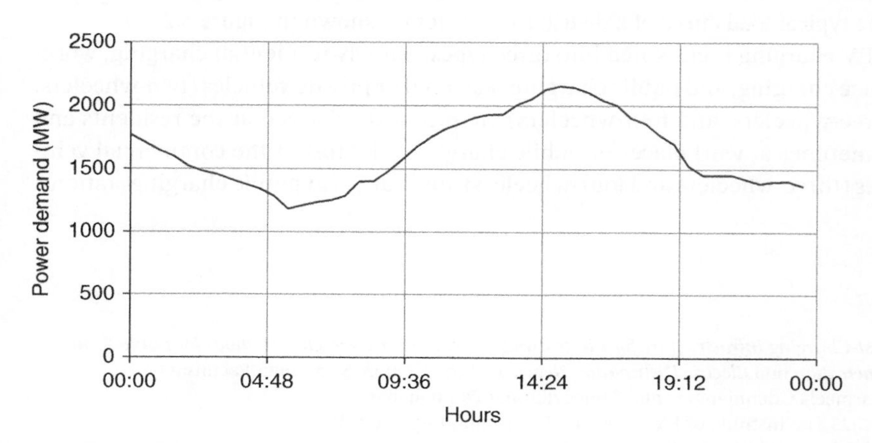

Figure 5.2 Typical load curve of a distribution system.

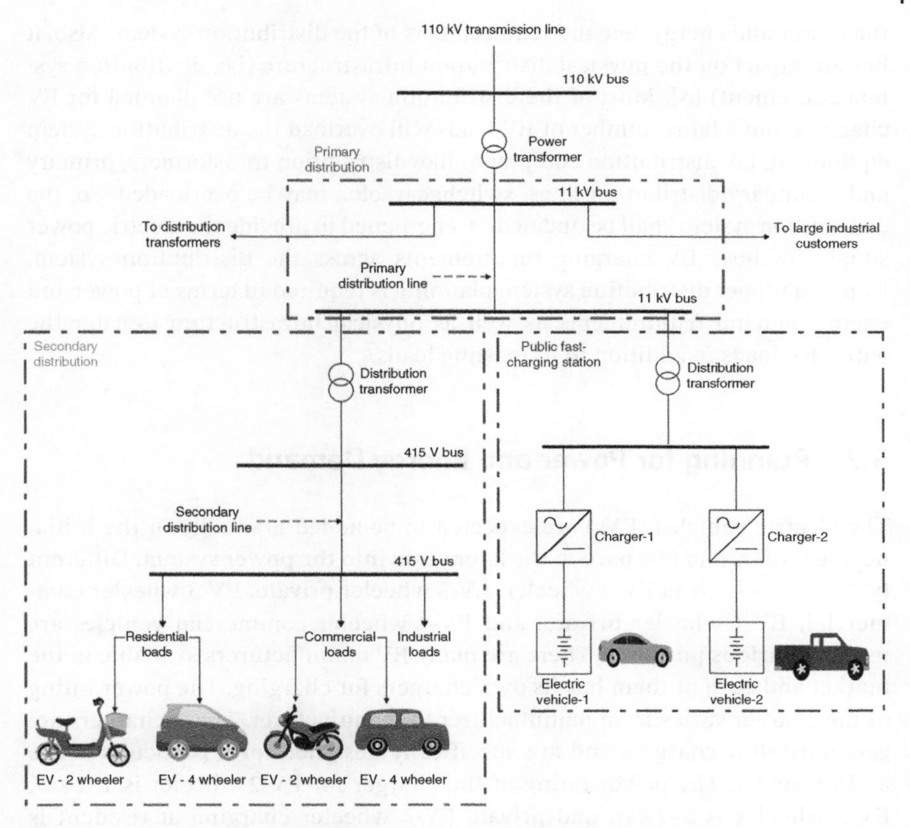

Figure 5.3 General structure of the distribution system with EV loads.

Hence, the EV charging loads are expected to connect and charge across the distribution. Figure 5.3 shows the typical structure of the distribution system with the inclusion of EV charging loads. Generally, the power rating of the charger for EV-2 wheeler is 1–2 kW, EV-3 wheeler is 3–7 kW, and private EV-4 wheeler charging at resident is 5–7 kW. The chargers used for EV charging in residences are slow chargers and it will take two to six hours for complete charging (from 0% to 100%). In order to reduce the time required for charging, DC fast chargers are widely used in public charging stations. These public charging stations are mostly located on highways, cinema theaters, shopping malls, hospitals, etc. The general power rating of DC fast chargers is in the range of 50–150 kW and it will charge the EVs in lesser time duration as compared with slow chargers. The large penetration of EVs such as EV-2 wheelers, EV-3 wheelers, and EV-4 wheelers into the distribution system will increase the power and energy demand requirements of the distribution system. Also, it has an impact on the physical distribution infrastructure (i.e. distribution system equipment) [5]. Most of the distribution systems are not planned for EV charging and a large number of EV loads will overload the distribution system equipment, i.e. distribution equipment like distribution transformers, primary and secondary distribution lines, switchgear, etc., may be overloaded. So, the distribution system shall be updated/strengthened to provide the electric power supply to these EV charging requirements across the distribution system. Hence, a proper distribution system planning is required in terms of power and energy demand requirements as well as physical infrastructure to cater the entire EV loads in addition to its existing loads.

5.2 Planning for Power and Energy Demand

The electric vehicles (EVs) are expected to be added gradually in the initial next few years and fast pace in the later years into the power system. Different types of EVs such as EV-2 wheeler, EV-3 wheeler private, EV-3 wheeler commercial, EV-4 wheeler private, and EV-4 wheeler commercial vehicles are used for various purposes. There are many EV manufacturers available in the market and each of them has its own chargers for charging. The power rating of the charger varies from manufacturer to manufacturer. These chargers are generally slow chargers and are specifically designed for a particular make and/or model. The power rating of the charger for EV-2 wheeler is 1–2 kW, EV-3 wheeler is 3–7 kW, and private EV-4 wheeler charging at resident is 5–7 kW. The chargers used for EV charging in residences are slow chargers and it will take two to six hours for complete charging. In order to reduce the time required for charging, DC fast chargers are widely used in public charging stations. These public charging stations are mostly located on highways, cinema theaters, shopping malls, hospitals, etc. The general power rating of DC fast chargers is in the range of 50–150 kW and it will charge the EVs in lesser time duration [6, 7].

The large penetration of EVs such as EV-2 wheelers, EV-3 wheelers, and EV-4 wheelers into the distribution system will increase the power and energy demand requirements of the distribution system. In order to predict the expected power and energy demand requirements to cater to the entire EV charging loads, it is important to know how many EVs (including their type) are expected to fly on road and their growth rate.

Example 5.1 Power and Energy Demand Estimation for EV Loads

Tamil Nadu is one of the states in India and located in the southern part of the country. As per the 2011 population census, 72 147 030 people are living in the state. It has more number of commercial and industrial establishments across the state. Tamil Nadu Generation and Distribution Company (TANGEDCO) is a DISCOM in Tamil Nadu and it is providing the electric power supply to its consumers across the state.

The EVs become an absolute necessity for transportation due to lesser operating cost and free from pollution. The Government of India is aiming to increase EV penetration into the market and they have a target for this. EV charging loads will increase the power and energy demand requirements of the TANGEDCO distribution system for Tamil Nadu, and they have to be well prepared to cater to all EV loads. Table 5.1 shows the total number of conventional vehicles registered in Tamil Nadu between the year 2014 and 2019 (the years 2020, 2021, and 2022 are not considered due to COVID impact) and growth rate in compound annual growth rate (CAGR).

Considering the vehicle registered in 2019 as base and CAGR from Table 5.1, the total number of conventional vehicles projection has been done for the year 2023–2030 and listed in Table 5.2.

Table 5.1 Total number of conventional vehicles registered in Tamil Nadu.

Type of vehicles	Year						CAGR (%)
	2014	2015	2016	2017	2018	2019	
4W private vehicles	14 757	17 907	23 648	20 281	21 394	18 513	4.6%
4W heavy vehicles	17 443	21 896	20 774	17 339	20 774	21 409	4.2%
4W commercial vehicles	204 151	203 794	249 263	264 971	287 100	266 234	5.5%
3W commercial vehicles	21 621	17 614	24 732	25 154	37 057	42 788	14.6%
2W private vehicles	1 417 935	1 377 214	1 592 414	1 584 916	1 748 330	1 619 672	2.7%

Table 5.2 Total number of vehicle projections from 2023 to 2030.

Type of vehicles	Year							
	2023	2024	2025	2026	2027	2028	2029	2030
4W private vehicles	22 195	23 225	24 303	25 430	26 610	27 844	29 136	30 488
4W heavy vehicles	25 222	26 277	27 376	28 521	29 714	30 956	32 251	33 600
4W commercial vehicles	329 240	347 197	366 132	386 100	407 158	429 364	452 780	477 474
3W commercial vehicles	73 872	84 678	97 064	111 262	127 537	146 193	167 577	192 090
2W private vehicles	1 801 539	1 850 111	1 899 993	1 951 219	2 003 827	2 057 853	2 113 336	2 170 315

The EV penetration rate target by Government of India is listed in Table 5.3.

From the total number of conventional vehicles from Table 5.2 and EV penetration target 5.3, the total number of EVs is estimated and listed in Table 5.4.

Following assumptions are considered to estimate the power and energy demand requirements of EVs listed in Table 5.4:

1) e-2W noncommercial (private) vehicles
   - All e-2-wheelers in the state are charged using AC slow charger only at home.
   - Each vehicle will charge only once in every three days. Hence, only 33.33% of total vehicles will be charged every day.
   - In the total number of vehicles, 30% of vehicles are charging at the residence in the morning and plug-in starts from 6 a.m. to 9 a.m. with an equal percentage share per hour and 70% of vehicles are charging in the evening hours and plug-in starts from 6 p.m. to 10 p.m. with an equal percentage share per hour.
   - At the time of plug-in, battery SoC is 20%, 30%, and 40%.
   - Among the total number of vehicles, equal share of vehicles are considered for 20%, 30%, and 40% SoC in both morning and evening.

Table 5.3 EV penetration rate target by Government of India (Adapted from https://rmi.org/wp-content/uploads/2019/04/rmi-niti-ev-report.pdf).

Types of vehicles	2023	2025	2030
e-2W penetration range	17%	32%	80%
e-3W commercial vehicle penetration range	33%	45%	80%
e-4W private vehicle penetration range	3%	5%	30%
e-4W commercial vehicle penetration range	13%	23%	70%
e-4W heavy vehicle penetration range	7%	11%	40%

Table 5.4 Total number of electric vehicles estimation for the year 2023, 2025, and 2030.

Types of vehicles	2023	2025	2030
Number of electric 2W (on-road)	306 262	607 998	17 36 252
Number of electric 3W commercial vehicles (on-road)	24 378	43 679	1 53 672
Number of electric 4W (on-road)-private vehicles	666	1215	9146
Number of electric 4W (on-road)-commercial vehicles	41 155	84 210	334 232
Number of electric 4W heavy vehicles including bus (on-road)	1766	3011	13 440

- Power rating of the charger at home (AC slow charger) is 1.7 kW.
- Battery capacity is 3 kWh for 2023, 4 kWh for 2025, and 6 kWh for 2030.
- C-rating of the battery is 0.5 for 2023, 0.6 for 2025, and 0.7 for 2030.

2) e-3W commercial vehicles
   - Commercial e-3Ws Autos in the state will be charged at the charging hub and public charging station.
   - In the total number of vehicles, 70% of vehicles are charging at the charging hub, and 30% of vehicles are charging at public charging.
   - At the time of plug, battery SoC is 20%, 30%, and 40%.
   - Among the total number of vehicles for each charging type, equal share of vehicles are considered for 20%, 30%, and 40% SoC.
   - Power rating of the charger at the charging hub (AC slow charger) is 7 kW.
   - Power rating of the charger at the public charging station (DC fast charger) is 20 kW.
   - Battery capacity is 8 kWh for 2023, 9 kWh for 2025, and 10 kWh for 2030.
   - C-rating of the battery is 0.5 for 2023, 0.6 for 2025, and 0.7 for 2030.
   - Charging hub: Plug-in starting from 12 p.m. in the afternoon to 11 p.m. in the night with equal % of share.
   - Public charging (mid-day charging): Plug-in starting from 12 p.m., 1 p.m., 2 p.m., 3 p.m., and 4 p.m. with equal % of share.
3) e-4W commercial vehicles
   - Commercial e-4Ws in the state will be charged at the charging hub and public charging station.
   - In the total number of vehicles, 50% of vehicles are charging at public charging and 50% of vehicles are charging at charging hubs.
   - At the time of plugin, battery SoC is 20%, 30%, and 40%.
   - Among the total number of vehicles for each charging type, equal share of vehicles are considered for 20%, 30%, and 40% SoC.
   - Power rating of the charger at the charging hub (AC slow charger) is 3.3 and 7 kW. Out of 100% of vehicle charging at charging hub, 50% of vehicles are charged by using 3.3 kW charger and remaining 50% of vehicles are charged by using 7 kW charger.
   - Power rating of the charger at the public charging station (DC fast charger) is 20 and 50 kW. Out of 100% of vehicle charging at public charging station, 50% of vehicles are charged by using 20 kW charger and remaining 50% of vehicles are charged by using 50 kW charger.
   - Battery capacity is 20 kWh for 2023, 25 kWh for 2025, and 35 kWh for 2030.
   - C-rating of the battery is 0.5 for 2023, 0.6 for 2025, and 0.7 for 2030.
   - Charging hub (overnight charging): Plug-in starting from 9 p.m., 10 p.m., and 11 p.m. % of the share is 33%.
   - Public charging (mid-day charging): Plug-in starting from 12 p.m., 1 p.m., 2 p.m., 3 p.m., and 4 p.m. % of the share is 20%.

4) e-4W noncommercial (private) vehicles
   - Noncommercial e-4Ws in a city will be charged at home, workplace charging, and public charging stations.
   - Each vehicle will only charge once in three days. Hence, only 33% of total vehicles will be charged every day.
   - In the total number of vehicles, 70% of vehicles are charged at residential charging, 10% of vehicles are charged at public charging stations, and 20% of vehicles are charged at the workplace.
   - At the time of charging, battery SoC is 20%, 30%, and 40%.
   - Among the total number of vehicles for each charging type, equal share of vehicles are considered for 20%, 30%, and 40% SoC.
   - Power rating of the charger at the charging hub (AC slow charger) is 7 kW.
   - Power rating of the charger at the public charging station (DC fast charger) is 50 kW.
   - Battery capacity is 30 kWh for 2023, 50 kWh for 2025, and 62 kWh for 2030.
   - C-rating of the battery is 0.5 for 2023, 0.6 for 2025, and 0.7 for 2030.
   - Residential charging: Plug-in starting from morning 6 a.m. to 9 a.m. and evening 7 p.m. to 10 p.m. % of the share is 20% for morning and 80% for evening. Among 20% of vehicles charging in the morning, 5% of vehicles are connected to the charger in every hour. Among 80% of vehicles charging in the evening, 20% of vehicles are connected to the charger in every hour.
   - Public charging: All the vehicles are using public charging between 11 a.m. and 6 p.m. and charging up to 80% of SoC. % of the share is 11 a.m. to 6 p.m. – 12.5% per hour.
   - Workplace charging (mid-day charging): Plug-in starting from morning 9 a.m. to 2 p.m.
5) e-heavy vehicles
   - All e-heavy vehicles in the state will be charged at the bus depot and public fast-charging stations dedicated for e-heavy vehicles only.
   - At the time of charging, battery SoC is 20%, 30%, and 40% in the night and 50% at the mid-day charging.
   - Among the total number of e-buses and e-heavy vehicles, an equal share of vehicles is considered for 20%, 30%, and 40% SoC.
   - Power rating of the charger at the e-bus depot and public fast-charging station is 180 kW DC fast charger.
   - Battery capacity is 200 kWh.
   - C-rating of the battery is 0.75.
   - Overnight charging: Plug-in starting from 11 p.m. to next day 3 a.m. with equal percentage of share in every hour.
   - Mid-day charging: Plug in starting from 12 p.m. to 4 p.m. with equal percentage of share in every hour.

6) Others
   - The efficiency of the charger is 95% for 3.3, 3.5, 7 kW AC slow charger, and 20 kW DC fast charger, 96% for 50 kW and above DC fast charger.
   - TANGEDCO distribution system loss is 15.14%.

The power demand in 24 hours to cater all EV loads for the year 2023, 2025, and 2030 is shown in Figures 5.4–5.6, respectively.

Table 5.5 shows the 24 hours power demand requirement to cater all EV charging loads for the year 2023, 2025, and 2030.

Table 5.6 shows the total daily and yearly energy requirements for the year 2023, 2025, and 2030.

The power and energy demand for charging electric vehicles for the state is listed in Tables 5.5 and 5.6. Both power and energy demand requirements are increasing year by year due to higher share of EV penetration. The electricity distribution company, i.e. TANGEDCO, has to plan for power and energy demand requirements to cater the EV loads that get connected and charged across the TANGEDCO distribution system across the state of Tamil Nadu. These EV loads are addition to the existing loads. This additional demand can be met by installing TANGEDCO-owned generating stations such as conventional power plants like thermal plants and nonconventional power plants like solar PV plants and wind power plants, plan for a long-term Power Purchase Agreement (PPA) with other generating companies.

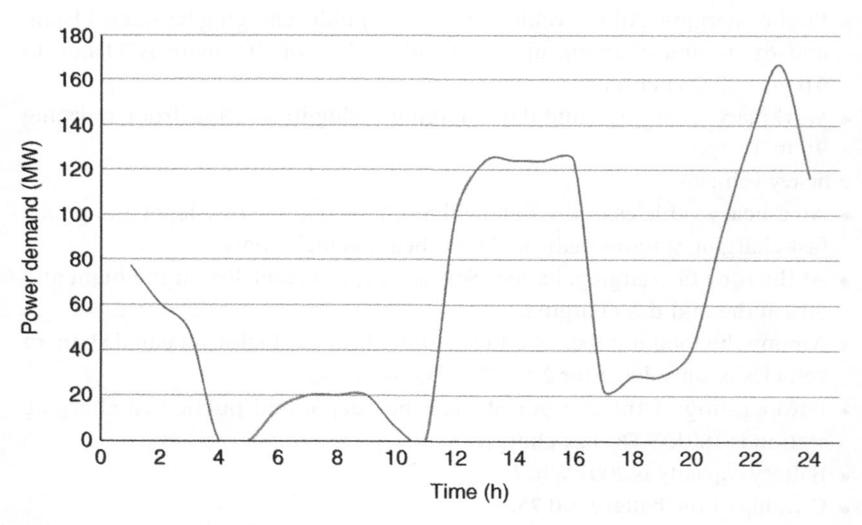

Figure 5.4 Power demand required for the year 2023.

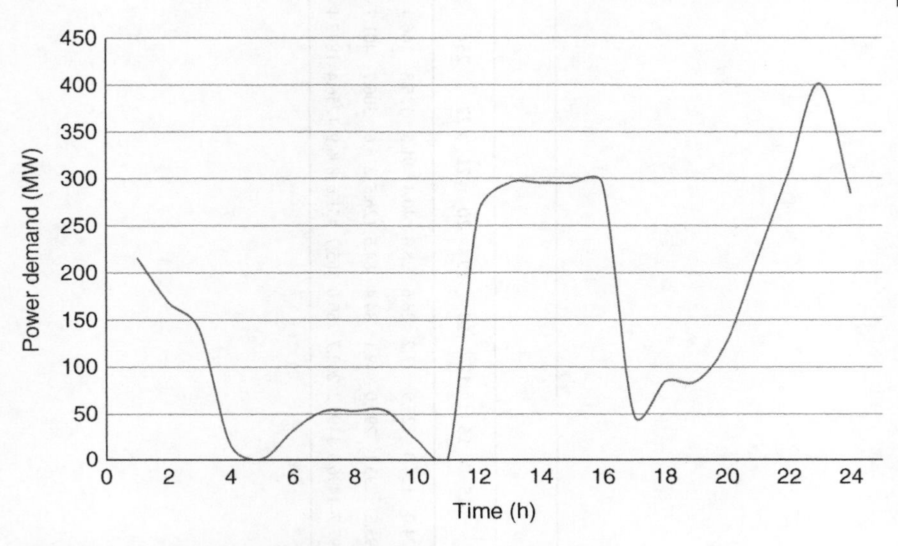

Figure 5.5 Power demand required for the year 2025.

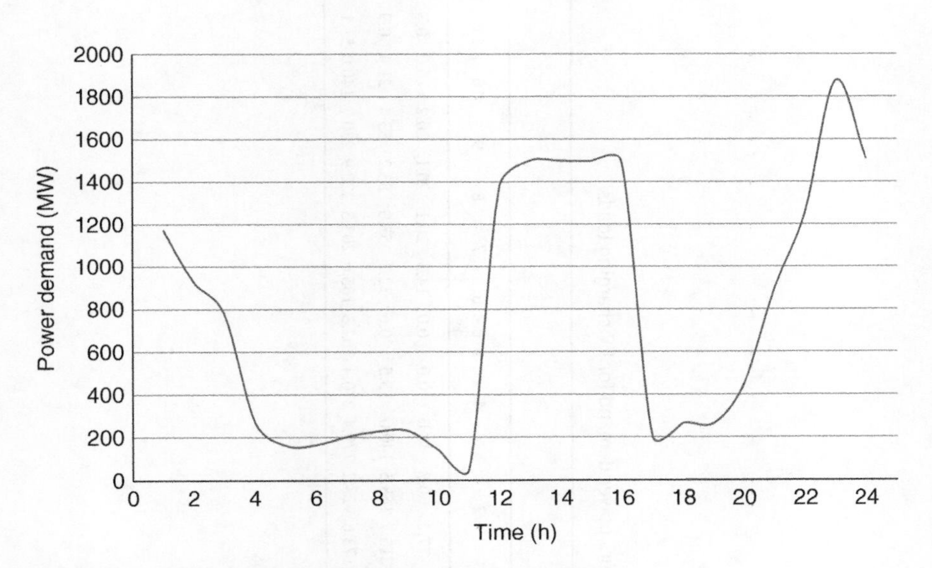

Figure 5.6 Power demand required for the year 2030.

Table 5.5 24 hours power demand for EV charging loads.

Parameter	Year	Time (h)																							
		1	2	3	4	5	6	7	8	9	10	11	12	13	14	15	16	17	18	19	20	21	22	23	24
Total power demand (MW)	2023	77.2	60.3	48.0	0.0	0.0	14.3	20.1	20.1	20.2	6.0	0.3	100.9	124.0	124.0	124.0	123.9	23.2	28.6	28.8	40.4	91.5	135.3	166.4	116.2
	2025	215.7	168.5	140.0	15.5	0.6	32.2	53.0	53.2	53.4	21.4	0.9	263.7	295.3	295.3	295.1	295.0	49.1	84.4	84.5	126.5	216.6	306.7	401.7	284.3
	2030	1173.6	928.2	779.4	270.4	162.3	165.0	203.3	228.9	230.7	140.0	51.1	1374.2	1497.9	1497.7	1496.5	1495.2	202.7	266.0	265.7	451.6	896.9	1256.4	1875.4	1507.5

Table 5.6 Total daily and yearly energy requirements.

S. No	Year	Total daily energy requirements (MWh)	Total yearly energy requirements (MWh)	Total yearly energy requirements (MU)
1	2023	1494	545 241	545
2	2025	3753	1 369 664	1370
3	2030	18 417	6 722 098	6722

5.3 Planning for Distribution System Feeders and Equipment

The EVs such as EV-2wheelers, EV-3wheelers, and EV-4wheelers are expected to connect and charge across the distribution system. The large penetration of EVs into the distribution system will increase the power and energy demand requirements of the distribution system. The required electric power for EV charging has to be provided by DISCOM through its distribution system in addition to the existing loads (i.e. without EV loads). The increased power flow across the distribution system has an impact on the distribution system equipment like overloading, improper or inefficient reactive power compensation, poor voltage profile in various buses, increased amount of current harmonic injection, etc. [8]. Present distribution system equipment are not planned for the large number of EV charging loads. So, the distribution system shall be upgraded/modified to provide the power supply to these EV charging requirements across the distribution system. Hence, proper planning of the distribution system is required in terms of power and energy demand requirements as well as physical distribution infrastructure to cater the entire EV loads [4].

The typical distribution infrastructure planning to cater the EV charging load is shown in Figure 5.7.

It is the responsibility of DISCOM to strengthen the distribution system infrastructure to cater all EV charging loads that are being connected and charged across the distribution system. This includes the modifications to the present distribution system equipment as well as new equipment and/or new feeders. A detailed power system analysis like load flow analysis, short-circuit analysis, harmonic analysis, etc., are to be performed to identify the present capability of the distribution system equipment without EV charging loads and additional requirements with the inclusion of EV charging loads.

Before starting the detailed power system analysis, it is much more important to know about the expected number of EVs and their types that may use the particular distribution system for charging. This information can be estimated from the

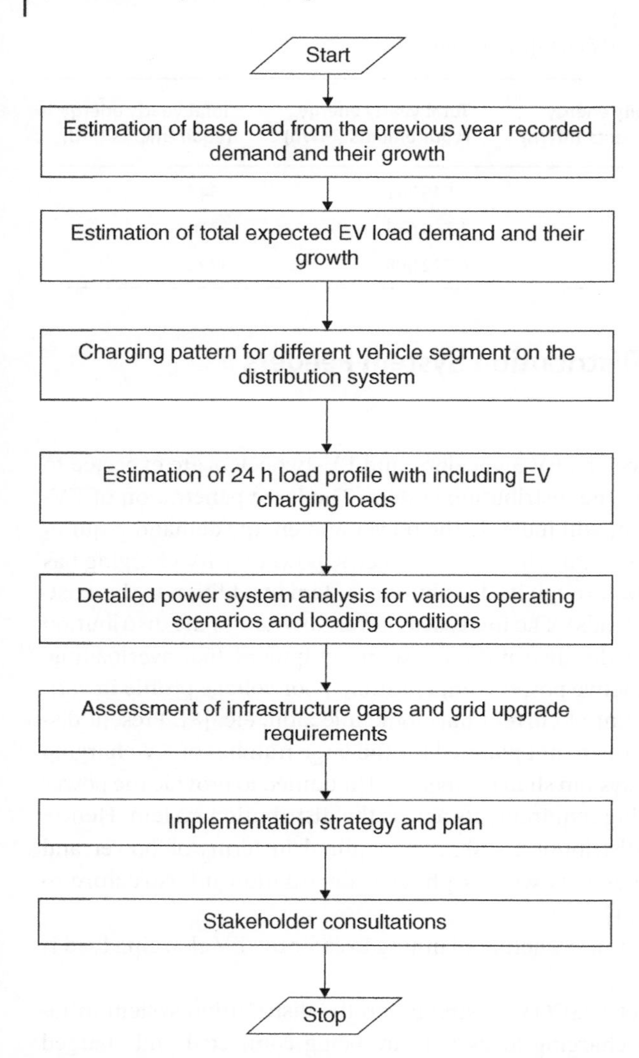

Figure 5.7 Typical distribution infrastructure planning to cater the EV charging load.

previous years registered vehicles in the city, their growth, and EV penetration level. This information has to be mapped with a particular city for estimation of the number of EVs concentrated by region and the expected number of public charging station requirements.

The geospatial analysis is used to identify the expected number of EVs concentrated in the region based on various factors like population, location of office/ industries, etc. The outcome of the geospatial analysis is used to estimate the number of charging stations needed for charging the EVs in that region. This information (i.e. expected number of charging stations required in that region) has to be mapped with DISCOM distribution system infrastructure for identifying the location of charging stations and their rating for planning.

A detailed power system analysis including load flow analysis, short-circuit analysis, harmonic analysis, etc. [9, 10], is essentially to be conducted for several cases like existing loads without the inclusion of EVs, existing loads with the inclusion of EVs considering their growth rate for the next 5–10 years period.

The load flow analysis provides the details of [11]:

- Voltage profile in various buses in the distribution system.
- Voltage drops in the distribution system.
- Equipment and circuits loadings in the distribution system.
- Distribution transformer rating, impedance, and tap requirements in the distribution system.
- Real and reactive power flow in the various branches of the distribution system.
- Real and reactive power losses in the distribution system.
- Reactive power compensation devices rating and their location, type of compensation (fixed or adjustable) in the distribution system.
- Performance of the distribution system under minimum, maximum, and normal loading conditions.

The outcome of load flow analysis, short-circuit analysis, and harmonic analysis provides the details of additional distribution system infrastructure requirements to meet the future EV charging loads in the distribution system. Additional distribution system infrastructure are modifications in the existing distribution system equipment/feeders and new distribution system equipment/feeders.

The typical modifications in the distribution system infrastructure are required to cater the future EV charging loads:

- Upgradation of existing distribution transformer from present rating to higher capacity.
- New distribution transformer rating, tap requirements, and their location in the distribution.
- Upgradation of existing primary and/or secondary distribution line to higher capacity.
- Splitting the existing primary and/or secondary distribution line into two or more lines.
- New primary and/or secondary distribution lines.
- Upgradation of reactive power compensation devices to higher capacity.
- New reactive power compensation device rating and their location in the distribution.

Example 5.2 Distribution Grid Infrastructure Upgradation for EV Loads

The Bangalore Electricity Supply Company Limited (BESCOM) is a DISCOM and it provides the electric power supply to its customers (more than 12 million) in the city of Bengaluru and its nearby places in India. It has the plan to develop 162 charging stations across Bengaluru city. As part of this, BESCOM has already installed a total of 136 charging stations which include 32 numbers of DC fast chargers and 104 numbers of AC slow chargers within Bengaluru. BESCOM has carried out a detailed power system analysis for 12 numbers of its 11 kV feeders with the inclusion of rooftop solar PV systems and EV charging loads for the years 2020, 2022, and 2025 [12, 13].

- Upgradation of harmonic filters to higher capacity.
- New harmonic filter rating, tuned frequency, and their location.
- New distributed generation resources rating and their location of integration into the distribution system.

The following data has been considered for the detailed power system analysis:

- Single line diagram (SLD) for 11 kV distribution feeder.
- Peak load demand was recorded for an 11 kV feeder in the year 2020.
- Distribution transformer rating, voltage ratios, vector group, and impedance.
- Cable and conductor characteristics.
- Feeders' current-carrying capacity.
- EV charging station details.
- GIS coordinates for EV charging loads and 11 kV feeders.
- Rooftop solar PV system penetration existing as well as the future years 2022 and 2025.

The study methodology adopted for the detailed study is shown in Figure 5.8. The summary of the 12 feeders considered for the study is listed in Table 5.7.

As per CEA technical standards for connectivity to the grid, the harmonic generation from the charging station has to comply with the limits specified in IEEE Std 519-2014. The voltage and current harmonic distortion limits at the point of common coupling (PCC) are listed in Tables 5.8 and 5.9, respectively.

The summary of load flow analysis including total load, losses, number of overloaded equipment, and number of under-voltage bus/nodes for 11 kV F08 Byatha feeder is listed in Table 5.10.

The current waveform at the PCC in the year 2022 is shown in Figure 5.9, and corresponding current harmonic spectrum is shown in Figure 5.10.

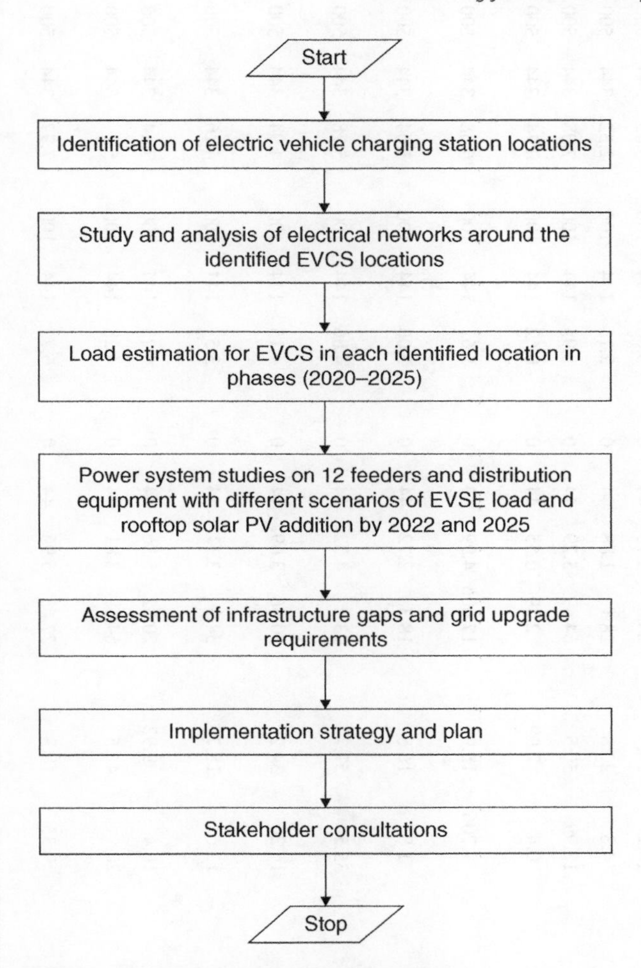

Figure 5.8 Flowchart of study methodology.

From Figures 5.9 and 5.10, it is observed that the current harmonic distortion at 3rd, 5th, 7th, and 11th orders exceeds the acceptable limits of IEEE 519. In order to mitigate the current harmonics injection into the system, it is recommended to install the harmonic filters at low voltage side. The required filter configuration to maintain the current harmonic limits with the acceptable limits are:

- Double-tuned filter configuration with rated capacitor power of 25 kVAR at 5th and 7th order.
- C-type filter rated 100 kVAR at 3rd order.

Table 5.7 Summary of the 12 feeders considered for the study [12, 13].

				Distribution line			2020			2022			2025		
S. No	Substation name	Feeder name	Number of distribution transformers	Length of overhead line (km)	Length of underground line (km)	Total length (km)	Peak load (MW)	EVSE load (kW)	Rooftop solar PV system (kW)	Peak load (MW)	EVSE load (kW)	Rooftop solar PV system (kW)	Peak load (MW)	EVSE load (kW)	Rooftop solar PV system (kW)
1	66/11 kV Arehally	F03 Ittamadu	22	10.1	18.5	28.6	2.73	44	0	3.067	144	100	3.65	344	500
2	66/11 kV HAL	F07 HAL	25	4.3	23.8	28.1	1.88	44	0	2.11	144	100	2.52	344	500
3	66/11 kV Hoody	F08 ITI AUX	61	13.99	50.5	64.49	5.29	44	0	5.95	144	100	7.08	344	500
4	66/11 kV Brindavana	F19 KIADB	9	0.4	2.09	2.49	0.55	44	0	0.62	144	100	0.74	344	500
5	66/11 kV Vidhana Soudha	F01 Cstation	11	1.706	15.9	17.606	4.99	44	0	5.61	144	100	6.68	344	500
6	66/11 kV Amarjyothi	F04 Cross-RMU	12	1.8	16.5	18.3	2.72	44	0	3.06	144	100	3.64	344	500
7	66/11 kV Banaswadi	F07 Subbaihanpalya	140	35.5	57.7	93.2	4.32	44	0	4.86	144	100	5.78	344	500
8	66/11 kV Banashankari	F25 BSK	55	16.26	34.2	50.46	3.69	44	0	4.15	144	100	4.94	344	500
9	220/11 kV HSR Layout	F03 HSR	64	15.6	18.7	34.3	3.11	44	0	3.5	144	100	4.16	344	500
10	66/11 kV KHB	F08 Byatha	37	10.4	9.92	20.32	5.56	44	0	6.25	144	100	7.44	344	500
11	66/11 kV Mathikere	F07 MTK	59	20	44.5	64.5	3.81	44	0	4.28	144	100	5.1	344	500
12	66/11 kV St. Johnswood	F01 SJW	66	9.35	18.3	27.65	5.96	44	0	6.7	144	100	7.97	344	500

Table 5.8 Voltage harmonic distortion limits at PCC *Source:* [14]/ with permission of IEEE.

Bus voltage V at PCC	Individual harmonic (%)	Total harmonic distortion THD (%)
$V \leq 1.0\,kV$	5.0	8.0
$1\,kV < V \leq 69\,kV$	3.0	5.0
$69\,kV < V \leq 161\,kV$	1.5	2.5
$161\,kV < V$	1.0	1.5[a]

[a] High-voltage systems can have up to 2.0% THD where the cause is an HVDC terminal whose effects will have attenuated at points in the network where future users may be connected.

Table 5.9 Current harmonic distortion limits for system 120V through 69kV at PCC *Source:* [14]/ with permission of IEEE.

	Maximum harmonic current distortion in percent of I_L					
	Individual harmonic order (odd harmonics)[a,b]					
I_{SC}/I_L	$3 \leqslant h < 11$	$11 \leqslant h < 17$	$17 \leqslant h < 23$	$23 \leqslant h < 35$	$35 \leqslant h \leqslant 50$	TDD
<20[c]	4.0	2.0	1.5	0.6	0.3	5.0
20 < 50	7.0	3.5	2.5	1.0	0.5	8.0
50 < 100	10.0	4.5	4.0	1.5	0.7	12.0
100 < 1000	12.0	5.5	5.0	2.0	1.0	15.0
>1000	15.0	7.0	6.0	2.5	1.4	20.0

[a] Even harmonics are limited to 25% of the odd harmonic limits above.
[b] Current distortions that result in a dc offset, e.g. half-wave converters, are not allowed.
[c] All power generation equipment is limited to these values of current distortion, regardless of actual I_{sc}/I_L.
where
I_{sc} maximum short-circuit current at PCC
I_L maximum demand load current (fundamental frequency component) at the PCC under normal load operating conditions

The mitigation measures/modifications proposed to strengthen the 11 kV F08 Byatha feeder to cater the EV loads planned for the year 2020, 2022, and 2025 are listed in Table 5.11.

In order to cater the EV charging loads for the years 2020, 2022, and 2025 for feeder F08 Byatha, BESCOM has to upgrade its F08 Byatha feeder by installing the new distribution transformers and harmonic filters as listed in Table 5.11.

Table 5.10 Summary of load flow analysis for 11 kV F08 Byatha feeder [12, 13].

Objective	Year 2020 without EV	Year 2020 with EV	Year 2022 with EV and solar PV	Year 2025 with EV and solar PV
Total source dispatch (kW)	5556	5602	6207	7018
Total load (kW)	4838	4841	5354	6182
Total loss (kW)	717	721	823	1005
Total EV load (kW)	0	44	144	344
Total PV generation (kW)	0	0	100	500
Total source (kVAR)	6185	2733	3026	3532
Total loss (kVAR)	1533	1548	1712	2001
No of overloaded elements	570	570	574	575
No of under-voltage bus/nodes	149	149	152	164

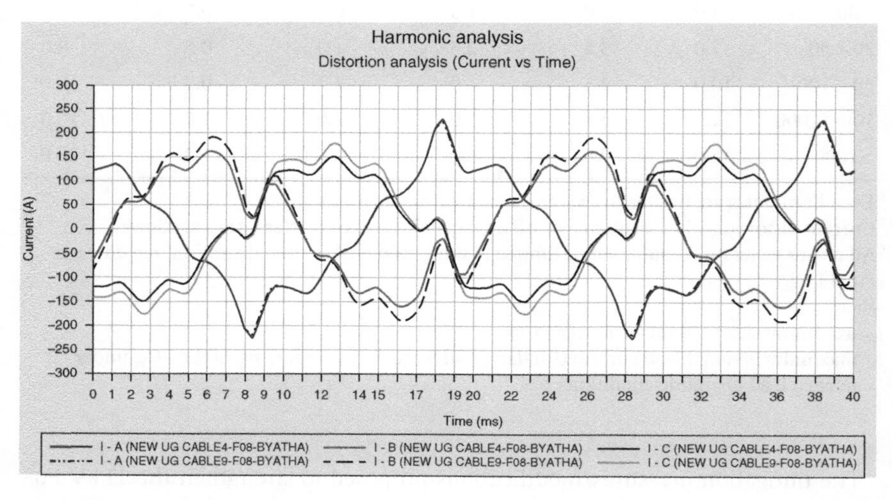

Figure 5.9 Current waveform at the PCC in the year 2022 [12, 13].

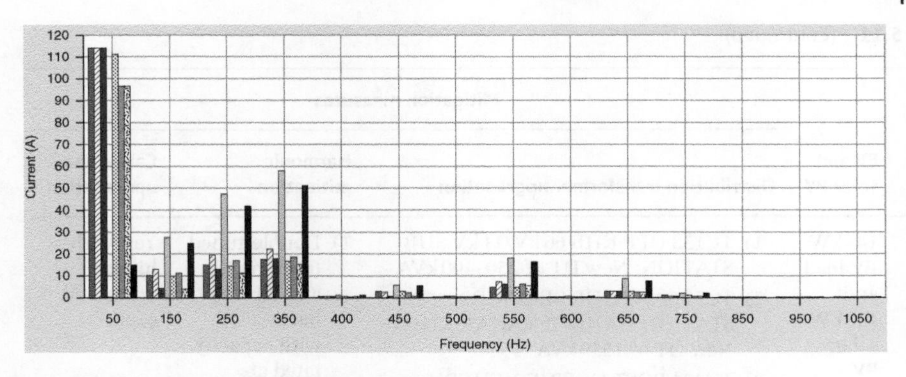

Figure 5.10 Current harmonic spectrum at the PCC in the year 2022 [12, 13].

Table 5.11 Mitigation measures/modifications proposed to strengthen the 11 kV F08 Byatha feeder [12, 13].

		Mitigation measures		
Year	**EV and solar PV**	**Distribution transformer upgradation**	**Harmonic mitigation**	**Cable/line upgradation**
2020	44 kW EV load	1) TC138 OPP KHB 66 kV/11 kV SUB STATION: New DT of 250300 kVA 2) TC52 SANTHUR SERVICE STATION GANDHI NAGAR KHB: New DT of 1500 kVA 3) TC384 POST OFFICE NEHERU NAGAR TC: New DT of 630 kVA 4) TC124 BUS STOP POST OFFICE YNK: New DT of 25 kVA 5) TC56 SHUSHRUSHA NURSHING HOME: New DT of 900 kVA 6) TC258 KENDRIYA VIHAR BHAVAN: New DT of 500–1000 kVA 7) TC257 KENDRIYA VIHAR BHAVAN: New DT of 500–1000 kVA 8) TC73 BAZAR ROAD: New DT of 250 kVA 9) TC49 NEAR ASWINI HOSPITAL: New DT of 160 kVA 10) TC135 OPP KHB 66 kV/11 kV SUB STATION: New DT of 250–350 kVA	1) Double-tuned filter for 5th and 7th harmonic with capacity rated at 15 kVAR 2) C-type filter rated at 30 kVAR tuned at 3rd order harmonic frequency	Yes, with higher withstand capacity

(*Continued*)

Table 5.11 (Continued)

Year	EV and solar PV	Mitigation measures		
		Distribution transformer upgradation	Harmonic mitigation	Cable/line upgradation
2022	144 kW EV load and 100 kW solar PV	1) TC138 OPP KHB 66 kV/11 kV SUB STATION: New DT of 250–400 kVA 2) TC52 SANTHUR SERVICE STATION GANDHI NAGAR KHB: New DT of 1550 kVA 3) TC384 POST OFFICE NEHERU NAGAR TC: New DT of 700 kVA 4) TC124 BUS STOP POST OFFICE YNK: New DT of 50 kVA 5) TC56 SHUSHRUSHA NURSHING HOME: New DT of 1000 kVA 6) TC258 KENDRIYA VIHAR BHAVAN: New DT of 500–1150 kVA 7) TC257 KENDRIYA VIHAR BHAVAN: New DT of 500–1000 kVA 8) TC73 BAZAR ROAD: New DT of 315 kVA 9) TC49 NEAR ASWINI HOSPITAL: New DT of 160 kVA 10) TC135 OPP KHB 66 kV/11 kV SUB STATION: New DT of 250–400 kVA	1) Double-tuned filter for 5th and 7th harmonic with capacity rated at 25 kVAR 2) C-type filter rated at 100 kVAR tuned at 3rd order	Yes, with higher withstand capacity
2025	344 kW EV load and 500 kW solar PV	1) TC138 OPP KHB 66 kV/11 kV SUB STATION: New DT of 250–400 kVA 2) TC52 SANTHUR SERVICE STATION GANDHI NAGAR KHB: New DT of 1550 kVA 3) TC384 POST OFFICE NEHERU NAGAR TC: New DT of 700 kVA 4) TC124 BUS STOP POST OFFICE YNK: New DT of 50 kVA 5) TC56 SHUSHRUSHA NURSHING HOME: New DT of 1000 kVA 6) TC258 KENDRIYA VIHAR BHAVAN: New DT of 500–1150 kVA 7) TC257 KENDRIYA VIHAR BHAVAN: New DT of 500–1000 kVA 8) TC73 BAZAR ROAD: New DT of 315 kVA 9) TC49 NEAR ASWINI HOSPITAL: New DT of 160 kVA 10) TC135 OPP KHB 66 kV/11 kV SUB STATION: New DT of 250–400 kVA	1) Double-tuned filter for 5th and 7th harmonic with capacity rated at 20 and 10 kVAR 2) C-type rated at 100 kVAR at 3rd order 3) Double-tuned filter of 75 and 30 kVAR C-type filter	Yes, with higher withstand capacity

5.4 Conclusion

The EVs are alternatives to conventional vehicles and they use energy stored in the battery for their operation. Whenever the EVs are running on the road, the energy stored in the battery is used to run the vehicles. After running on the road for some length (km), the energy stored in the battery is reduced. Recharging of the battery is required to extend the range of the vehicle. The EVSE is used to recharge the EVs whenever and wherever required. These EVs are expected to connect and charge across the distribution system. The rapid increase of EV charging will increase the power and energy demand requirements of the distribution system to cater the entire EV loads. Also, the physical distribution system shall be equipped to power these EV charging requirements across the distribution system. Hence, proper planning of the distribution system is required in terms of power and energy demand requirements as well as physical infrastructure to cater the entire EV loads. This chapter discussed the planning of the distribution system to meet the future EV charging loads across the distribution. The example of EV charging loads on peak power demand requirements for TANGEDCO DISCOM providing the power supply to Tamil Nadu state in India is explained in detail. Also, the example of distribution system upgradation required to cater the EV charging loads in Bengaluru is explained in detail.

References

1. Sivaraman, P. and Sharmeela, C. (2020). Introduction to electric distribution system. In: *Handbook of Research on New Solutions and Technologies in Electrical Distribution Networks* (ed. B. Kahn, H.H. Alhelou, and G. Hayek), 1–31. Hershey, PA: IGI Global.
2. IEEE Std 141-1993 (R1999) (1993). *IEEE Recommended Practice for Electric Power Distribution for Industrial Plants*. IEEE Standards.
3. El-Hawary, M.E. (2008). *Introduction to Electrical Power Systems*, 1e. Wiley-IEEE Press.
4. Sivaraman, P., Sharmeela, C., and Logeshkumar, S. (2021). Charging infrastructure layout and planning for plug-in electric vehicles. In: *Cable Based and Wireless Charging Systems for Electric Vehicles* (ed. P. Sanjeevikumar), 1–24. IET.
5. Bayram, I. and Tajer, A. (2017). *Plug-in Electric Vehicle Grid Integration*, 1e. Artech.
6. Dominguez-Navarro, J.A., Dufo-Lopez, R., Yusta-Loyo, J.M. et al. (2019). Design of an electric vehicle fast-charging station with integration of renewable energy and energy storage systems. *Electrical Power and Energy Systems* 105: 46–58.
7. Gnann, T., Funke, S., Jakobson, N. et al. (2018). Fast charging infrastructure for electric vehicles: today's situation and future needs. *Transportation Research Part D* 62: 314–329.

8 Sivaraman, P. and Sharmeela, C. (2020). Existing issues associated with electric distribution system. In: *Handbook of Research on New Solutions and Technologies in Electrical Distribution Networks* (ed. B. Kahn, H.H. Alhelou, and G. Hayek), 1–31. Hershey, PA: IGI Global.

9 Khan, S. (2008). *Industrial Power Systems*, 1e. CRC Press.

10 Saccomanno, F. (2003). *Electric Power Systems Analysis and Control*, 1e. Wiley-IEEE Press.

11 Sivaraman, P., Sharmeela, C., and Elango, S. (2021). Load flow analysis for microgrids. In: *Microgrid Technologies* (ed. C. Sharmeela and P. Sivaraman), 177–196. Scrivener Publishing LLC.

12 Anand Singh, Akshay Kumar, Balasubramanyam K, N. Murugesan, Reena Suri, Reji Kumar Pillai, and Shaurya Sharma ISGF Study of Electric Vehicle Charging Infrastructure Planning and Rollout for Bengaluru City, Karnataka, Volume 1, September 2021. Indian Smart Grid Forum.

13 Singh, A., Kumar, A., Balasubramanyam, K. et al. (2021). *ISGF study of Electric Vehicle Charging Infrastructure Planning and Rollout for Bengaluru City, Karnataka, Volume 2: Annexures*. Indian Smart Grid Forum.

14 IEEE Std 519-2014 (2014). *IEEE Standard for Harmonic Control in Electric Power Systems*. IEEE Standards.

6

Electric Distribution for Fast-Charging Infrastructure

6.1 Introduction

The electric vehicles (EVs) are an absolute necessity in this modern world because they are free from greenhouse gases, scarcity of fossil fuels, lesser fuel cost for operation, etc. These EVs are generally classified as battery-operated EVs (BEVs) and hybrid EVs (HEVs) [1]. Batteries are one of the important components in these vehicles. These batteries are similar to petrol and diesel in conventional vehicles and are used to run the electric motor during the vehicle operation. Like fuel capacity is limited to fuel tank capacity in a conventional vehicle, batteries in EVs are also limited based on their capacity in Ampere-Hour (Ah) or kiloWatt hour (kWh). As it is essential to refill the fuel in a conventional vehicle, it is critical to recharge the batteries in EVs. The components used to recharge the batteries in EVs are called electric vehicle supply equipment (EVSE) as per IEC [2], i.e. charging stations. It consists of chargers, connectors, distribution transformers, cables/lines, energy meters, etc. These charging stations are categorized based on the location of charging, charging speed, and type of charging current used. Based on the location of charging, it is categorized as a residential charging point (for small vehicles for personal use typically charging at home) and public charging point (for commercial vehicles and even mass charging of small passenger vehicles as in a Technology park or recreational Mall or on highways). Based on the speed of charging, it is categorized as slow charging and fast charging [3]. Based on the type of current used, it is categorized as DC charging and AC charging. AC charging: the conversion of AC/DC takes place inside the vehicle (on board). DC charging: the conversion of AC/DC takes place inside the EVSE (i.e. charging station). Figure 6.1 shows the typical AC charging and Figure 6.2 shows the typical DC charging used for EVs. However, the last

Fast-Charging Infrastructure for Electric and Hybrid Electric Vehicles: Methods for Large-Scale Penetration into Electric Distribution Networks, First Edition. Sivaraman Palanisamy, Sharmeela Chenniappan, and Sanjeevikumar Padmanaban.

Published 2023 by John Wiley & Sons, Inc.

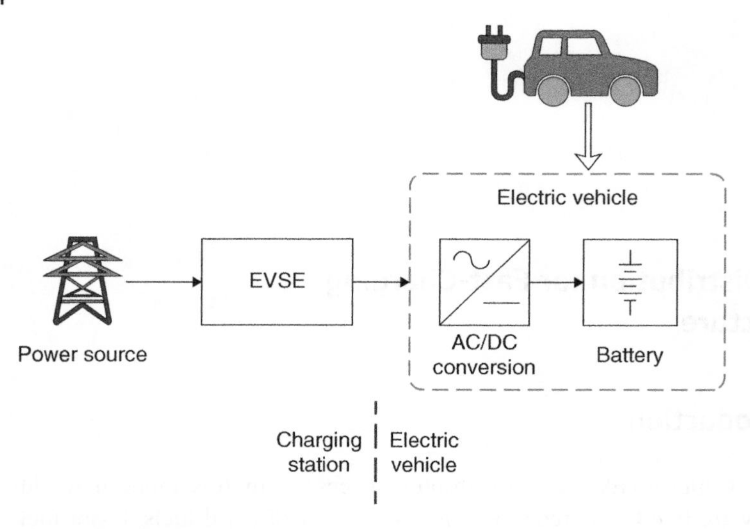

Figure 6.1 Typical AC charging.

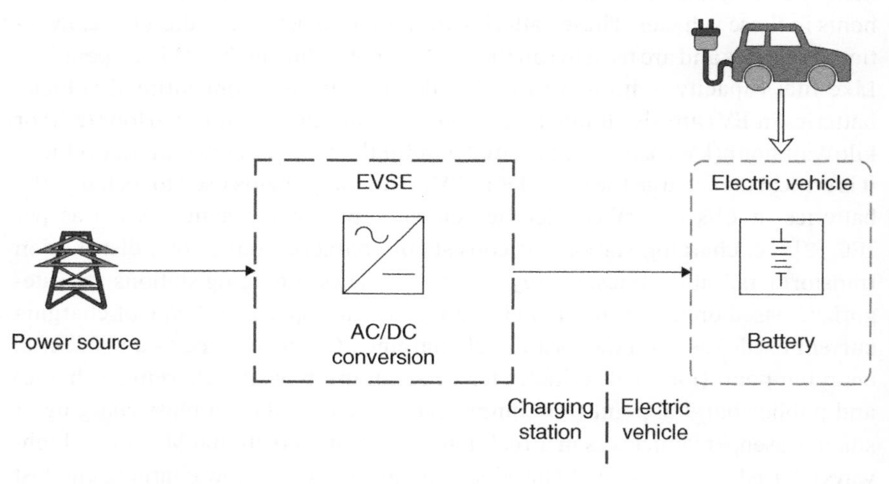

Figure 6.2 Typical DC charging.

classification would be better if standardized as AC charging or DC charging to move toward a one solution fits all system.

Most of the residential charging points are slow charging points and it will take more time for recharging the batteries. The typical time duration to recharge the battery is between two and six hours in slow charging. Slow charging is widely used to recharge the EVs used for personal use during the night-time [4].

Example: users will connect the EVs with chargers before they are going to bed and disconnect it the next morning. Similarly, most of the public charging points are fast-charging points and it will recharge the batteries faster in order to avoid the users' waiting time at the charging station [5].

These charging stations consist of electric equipment like distribution transformers, chargers, cables/lines, connectors, energy meters, ancillary equipment, etc. In simple words, they are called electrical distribution. A more hybrid setup for the charging stations is also possible and is discussed separately in Chapter 7. The load characteristics (i.e. power demand) of the charging station are varied with respect to time based on the SoC of the battery and the number of EVs recharging at a time. The typical load requirement for different SoC of the $LiFePO_4$ battery (battery voltage vs SoC and charging current vs SoC) is shown in Figure 6.3a,b (https://www.powerstream.com/LLLF.htm).

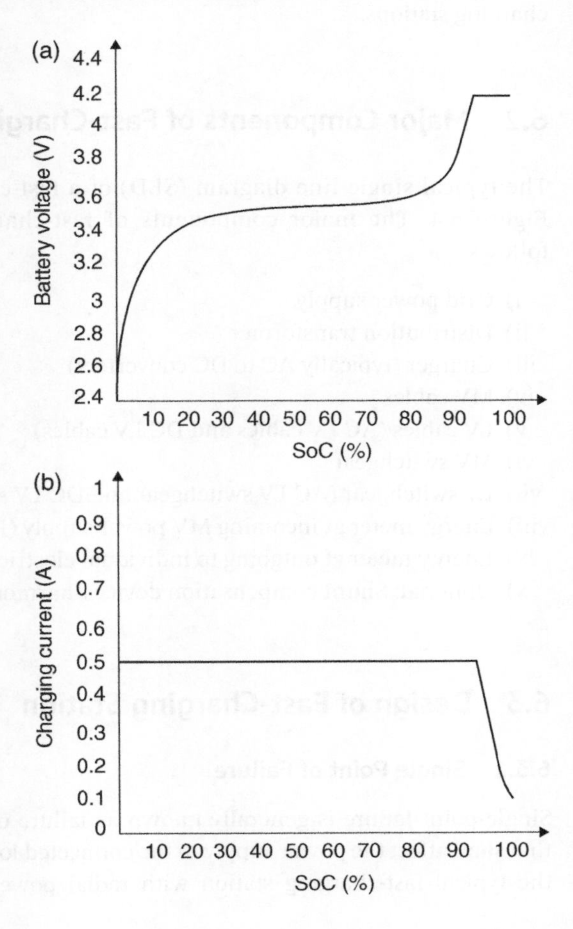

Figure 6.3 (a) Battery voltage vs SoC characteristics. (b) Charging current vs SoC characteristics.

Whenever the SoC of the battery is low, the constant current mode is used to recharge the battery. But, if the SoC of the battery is high (say more than 90%), constant voltage mode is used to recharge the battery, i.e. in a single recharging cycle, the current used to recharge the battery will vary with respect to SoC. The various reasons like electrical faults, failure of equipment, temporary outage, or tripping of equipment, etc. [6, 7], shall affect the recharging capability of the charging station leading to stoppage of charging at the station. Until the problem is resolved, the charging station will not able to recharge the EVs [8]. Reliability is a big question here. Hence, it is important to consider the reliability also in the design stage for selecting the electrical distribution [9, 10] for the fast-charging station. Contingency planning will also play a critical role in ensuring operation of a section of the charging station if not the entire station for certain outages. This would also enhance efficient planning and operation of a section of the charging stations.

6.2 Major Components of Fast-Charging Station

The typical single line diagram (SLD) of a fast-charging station is shown in Figure 6.4. The major components of fast-charging stations (FCS) are as follows:

i) Grid power supply
ii) Distribution transformer
iii) Charger (typically AC to DC converters)
iv) MV cables
v) LV cables (AC LV cables and DC LV cables)
vi) MV switchgear
vii) LV switchgear (AC LV switchgear and DC LV switchgear)
viii) Energy meter at incoming MV power supply (PCC)
ix) Energy meter at outgoing to individual electric vehicle
x) Optional: Shunt compensation devices/harmonic filters

6.3 Design of Fast-Charging Station

6.3.1 Single Point of Failure

Single-point failure is generally known as failure or outage of any components that shall affect the power supply to the connected loads [11, 12]. Figure 6.4 shows the typical fast-charging station with radial power distribution from the grid

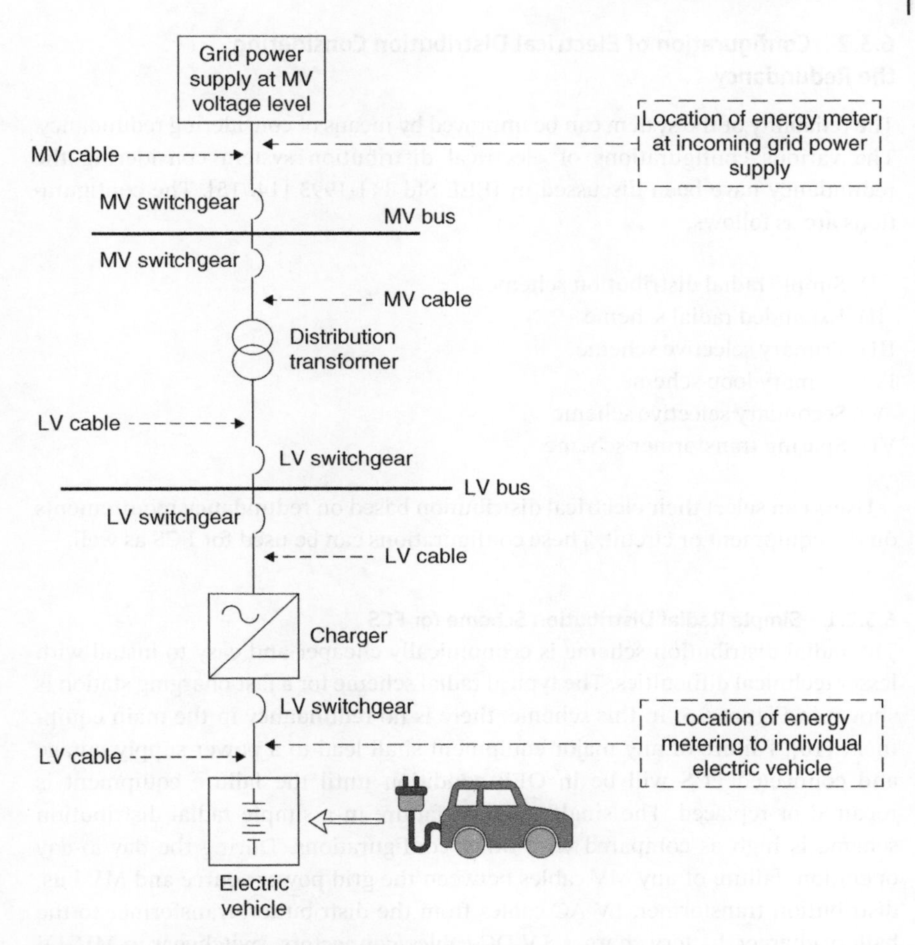

Figure 6.4 SLD for typical fast-charging station.

power supply at the MV voltage level to the electric vehicle charging point. In a fast-charging station with a radial distribution system, the chances of single-point failure are high. For example, in Figure 6.4, failure or outage of any of the MV cable, MV switchgear, LV switchgear, and distribution transformer shall lead to single-point failure and it will affect the EV charging. Till the fault is resolved or equipment has been replaced with a new one, the particular charging station is out of order and not in the position to recharge any EVs [13]. In the design and planning stage, necessary redundancy options are to be considered for achieving high reliability.

6.3.2 Configuration of Electrical Distribution Considering the Redundancy

The reliability of the system can be improved by means of considering redundancy. The various configurations of electrical distribution system considering the redundancy have been discussed in IEEE Std 141-1993 [14, 15]. The configurations are as follows,

I) Simple radial distribution scheme
II) Expanded radial scheme
III) Primary selective scheme
IV) Primary loop scheme
V) Secondary selective scheme
VI) Sparing transformer scheme

Users can select their electrical distribution based on redundancy requirements on the equipment or circuit. These configurations can be used for FCS as well.

6.3.2.1 Simple Radial Distribution Scheme for FCS

The radial distribution scheme is economically cheaper and easy to install with lesser technical difficulties. The typical radial scheme for a fast-charging station is shown in Figure 6.5. In this scheme, there is no redundancy in the main equipment [16]. Failure of any major equipment shall lead to a power supply outage and connected FCS will be in OFF condition until the failure equipment is repaired or replaced. The single point of failure in a simple radial distribution scheme is high as compared with other configurations. During the day-to-day operation, failure of any MV cables between the grid power source and MV bus, distribution transformer, LV AC cables from the distribution transformer to the battery charger, battery charger, LV DC cables/connectors, switchgear in MV, LV AC, and LV DC shall result in a single point of failure.

The advantages of using a simple radial distribution scheme for FCS are:

1) Economically cheaper
2) Technical complexity is less

The disadvantages of using a radial distribution scheme for FCS are:

1) High chance of single point of failure resulting in a power outage in a FCS.
2) Until the fault is resolved or failure equipment is replaced, EVs will not be able to recharge.

This scheme is suitable in locations where FCS load is less and reliability is not an issue.

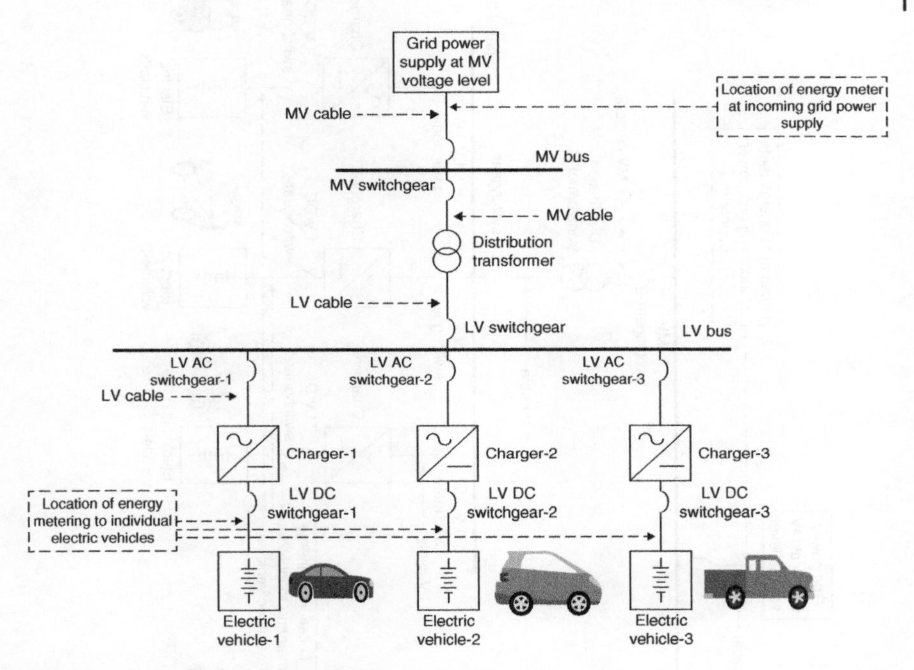

Figure 6.5 SLD for the radial scheme for FCS.

6.3.2.2 Expanded Radial Scheme for FCS

The expanded radial scheme is similar to the simple radial distribution scheme. In this scheme, the single grid power source is used to power the FCS loads through multiple distribution transformers. The typical expanded radial scheme for a fast-charging station is shown in Figure 6.6. The basic difference between a simple radial and expanded radial distribution schemes is listed in Table 6.1. This scheme is suitable for FCS where the EVs load requirement is high and reliability is not an issue.

6.3.2.3 Primary Selective Scheme for FCS

In this scheme, more than one grid power source is used to power the FCS loads as part of redundancy in the incoming grid power supply. The typical primary selective scheme for a fast-charging station is shown in Figure 6.7.

At a time, only one grid power source is used to power the loads. In case of failure of one grid power source, the other grid power source is used to power the FCS loads.

The operation of FCS loads connected through grid power source-1 is shown in Figure 6.8.

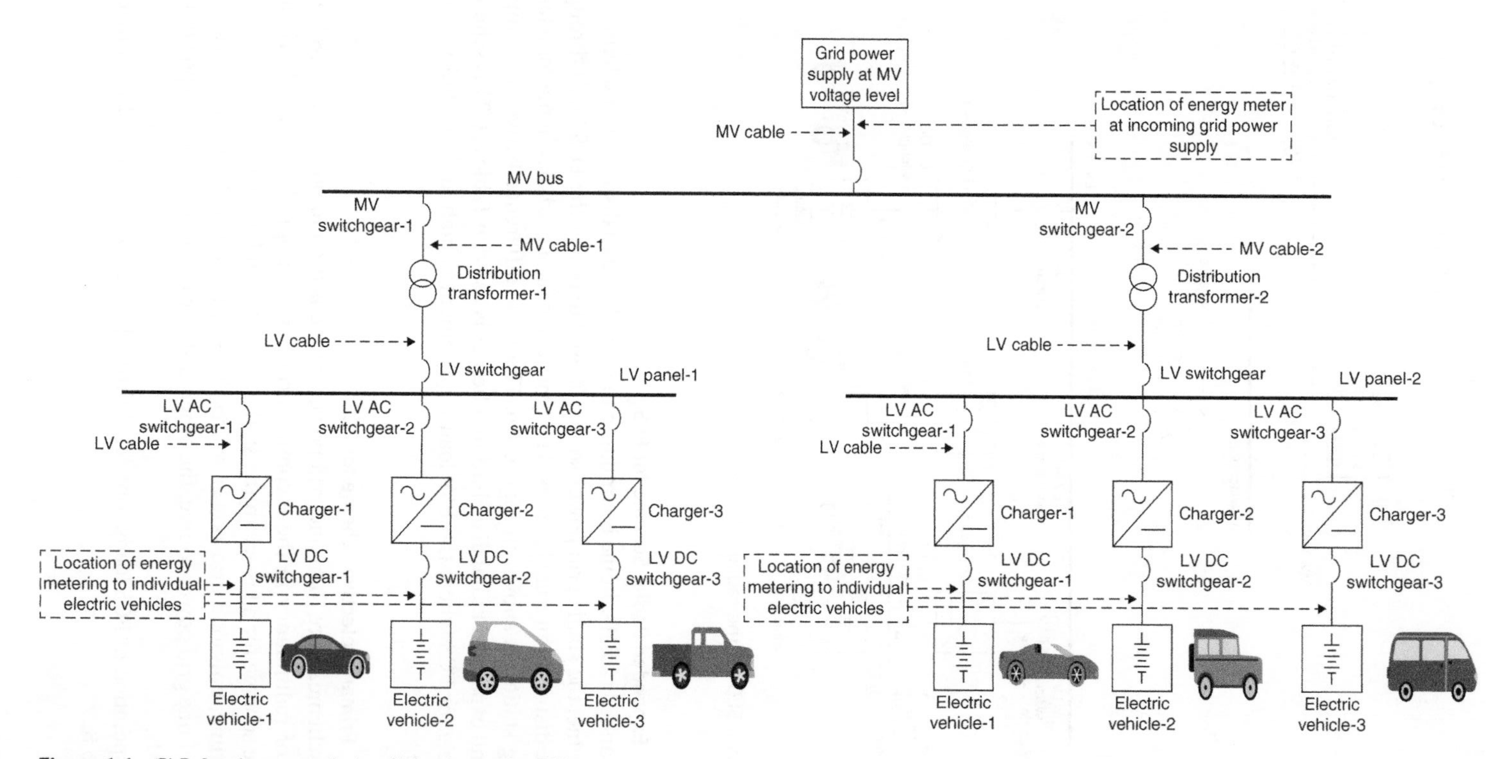

Figure 6.6 SLD for the expanded radial scheme for FCS.

Table 6.1 Difference between radial and expanded radial distribution scheme.

S. No	Expanded radial scheme	Simple radial scheme
1	Single grid power is used to power more than one distribution transformer	Single grid power is used to power one distribution transformer
2	The location of all distribution transformers is the same (same site premises) or different (different site premises) based on actual site conditions. Ex 1: The FCS has four numbers of distribution transformers and all transformers are installed adjacent to each other Ex 2: The FCS has four numbers of distribution transformers and all transformers are installed in different locations within the FCS	The location is not changed
3	Outage of any component under a distribution transformer will only lead to partial nonavailability of the FCS typically under that specific distribution transformer	Outage of any component could lead to complete disconnection of the FCS

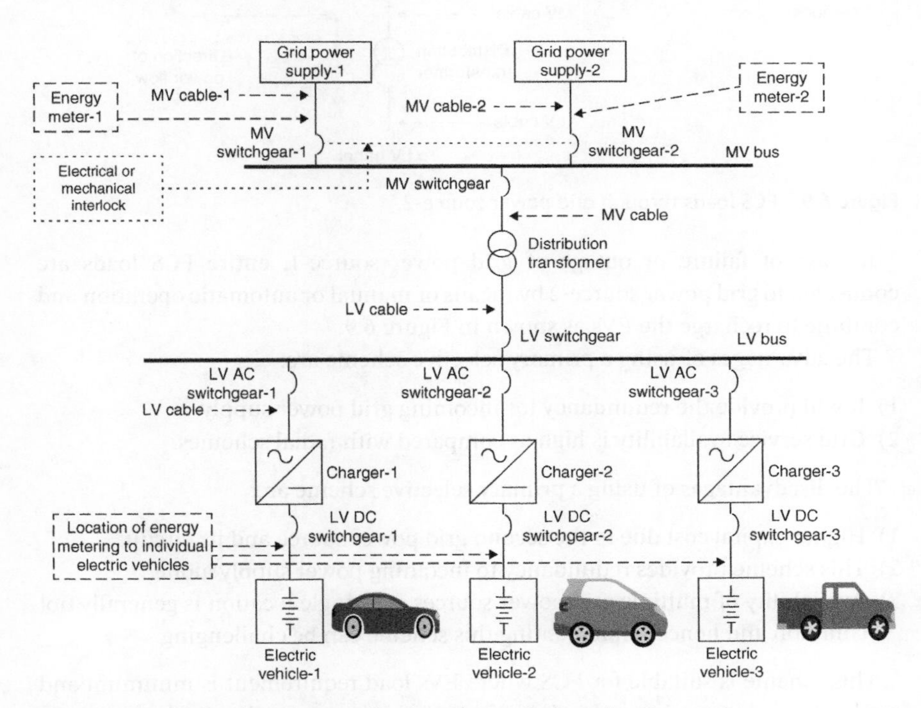

Figure 6.7 SLD for the primary selective scheme for fast-charging station.

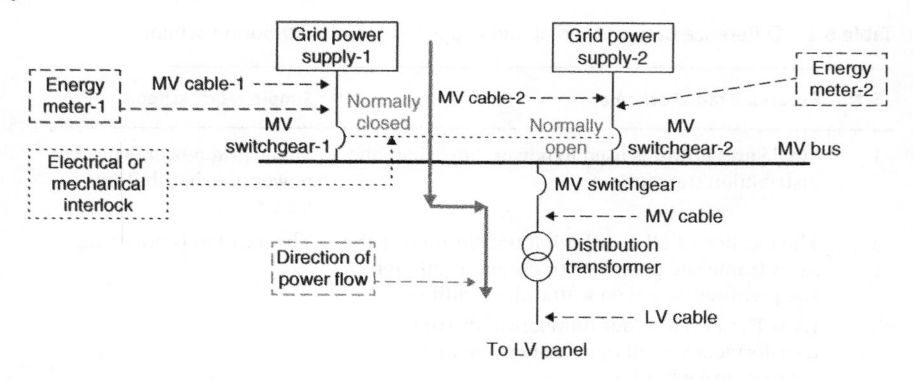

Figure 6.8 FCS loads through grid power source-1.

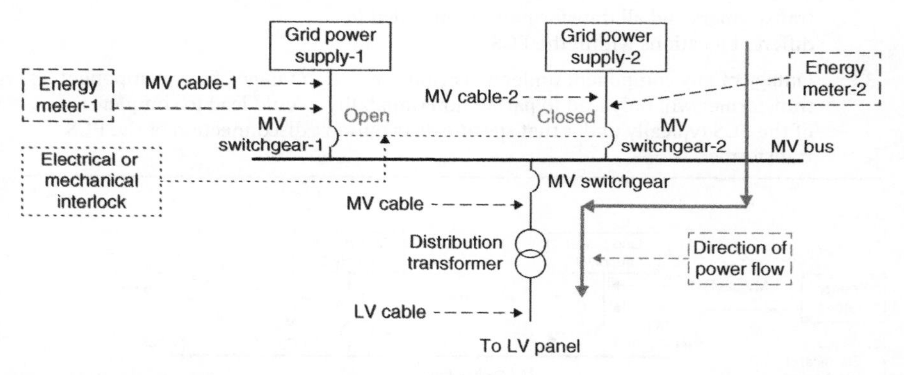

Figure 6.9 FCS loads through grid power source-2.

In case of failure or outage of grid power source-1, entire FCS loads are connected to grid power source-2 by means of manual or automatic operation and continue to recharge the EVs as shown in Figure 6.9.

The advantages of using a primary selective scheme are:

1) It will provide the redundancy for incoming grid power supply.
2) Grid service availability is high as compared with radial schemes.

The disadvantages of using a primary selective scheme are:

1) Higher capital cost due to the second grid power source and its circuit.
2) This scheme provides redundancy to incoming power supply alone.
3) Availability of multiple grid power sources at a single location is generally not common and hence implementing this scheme can be challenging.

This scheme is suitable for FCS where EVs load requirement is minimum and moderate, and it requires redundancy in the grid power supply.

6.3.2.4 Secondary Selective Scheme for FCS

This scheme involves one or more grid power sources, more than one distribution transformer connected to the same or different LV panel. The typical secondary selective scheme is explained in configuration 1 as shown in Figure 6.10 and configuration 2 as shown in Figure 6.11.

Configuration 1: One grid power supply at MV level
Configuration 2: Two independent grid power supplies at MV level

The secondary selective operation sequence of both configuration 1 and configuration 2 is:

- In normal operation, both distribution transformer-1 and distribution transformer-2 are in service (normally closed condition) and the bus coupler or tie-breaker is in open condition (normally open). Both distribution transformer-1 and distribution transformer-2 will equally share the FCS loads.
- In case of failure of distribution transformer-1 or its associated circuit, entire FCS loads connected to LV panel-1 powered by distribution transformer-1 will be connected to distribution transformer-2 either by means of manual or automatic operation of switching ON the bus coupler or tie-breaker as shown in Figure 6.12.
- Now, entire FCS loads are powered through distribution transformer-2.
- Once the fault is rectified or equipment replaced, then the FCS loads are transferred back to distribution transformer-1.
- Now both distribution transformer-1 and distribution transformer-2 continue to operate with equal loads.
- Similarly, failure of distribution transformer-2 and its circuits, distribution transformer-1 will continue to take care of FCS loads connected to distribution transformer-2 as shown in Figure 6.13.

Advantages of using a secondary selective scheme are:

- The redundant power supply is provided by another distribution transformer. Hence, reliability is increased.
- Redundancy is increased in configuration 2 with two incoming power supply.

Disadvantages of using a secondary selective scheme are:

- In this scheme, a minimum of two distribution transformers and their circuits are required. Hence, the capital cost is increased.
- To accommodate the load of either of the distribution transformer, the sizing of the transformer has to be doubled.

This scheme is more suitable for FCS where the load requirement is moderate and high with higher reliability.

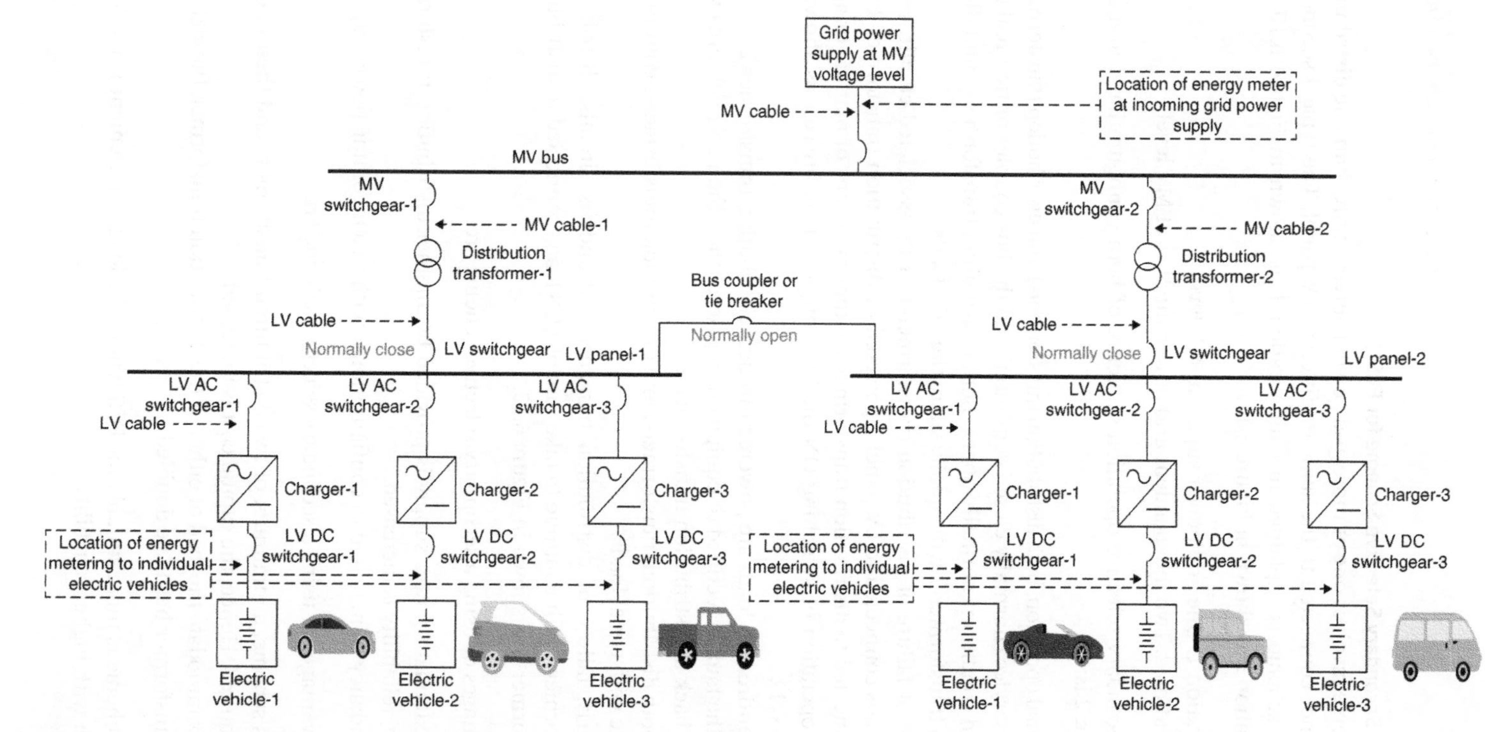

Figure 6.10 SLD for secondary selective scheme for FCS with one grid power supply.

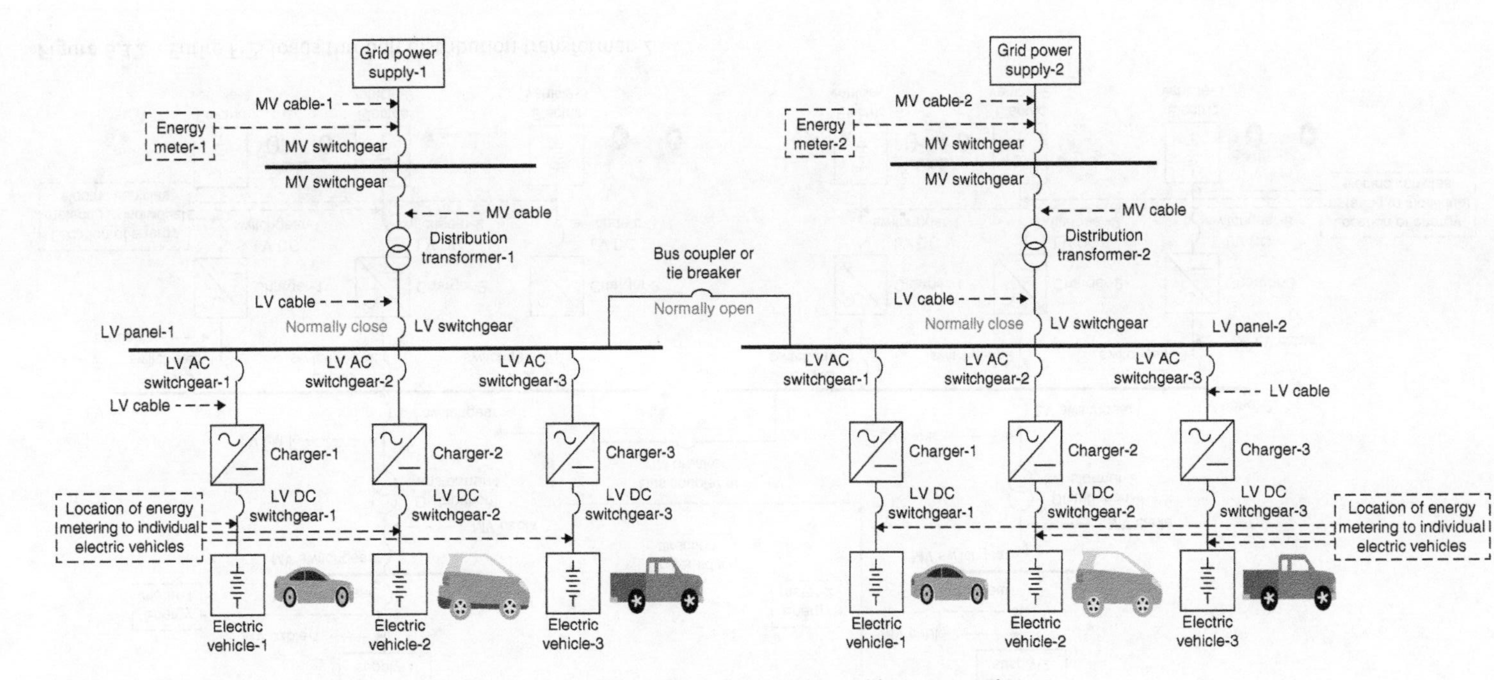

Figure 6.11 SLD for the secondary selective scheme for FCS with two independent grid power supply.

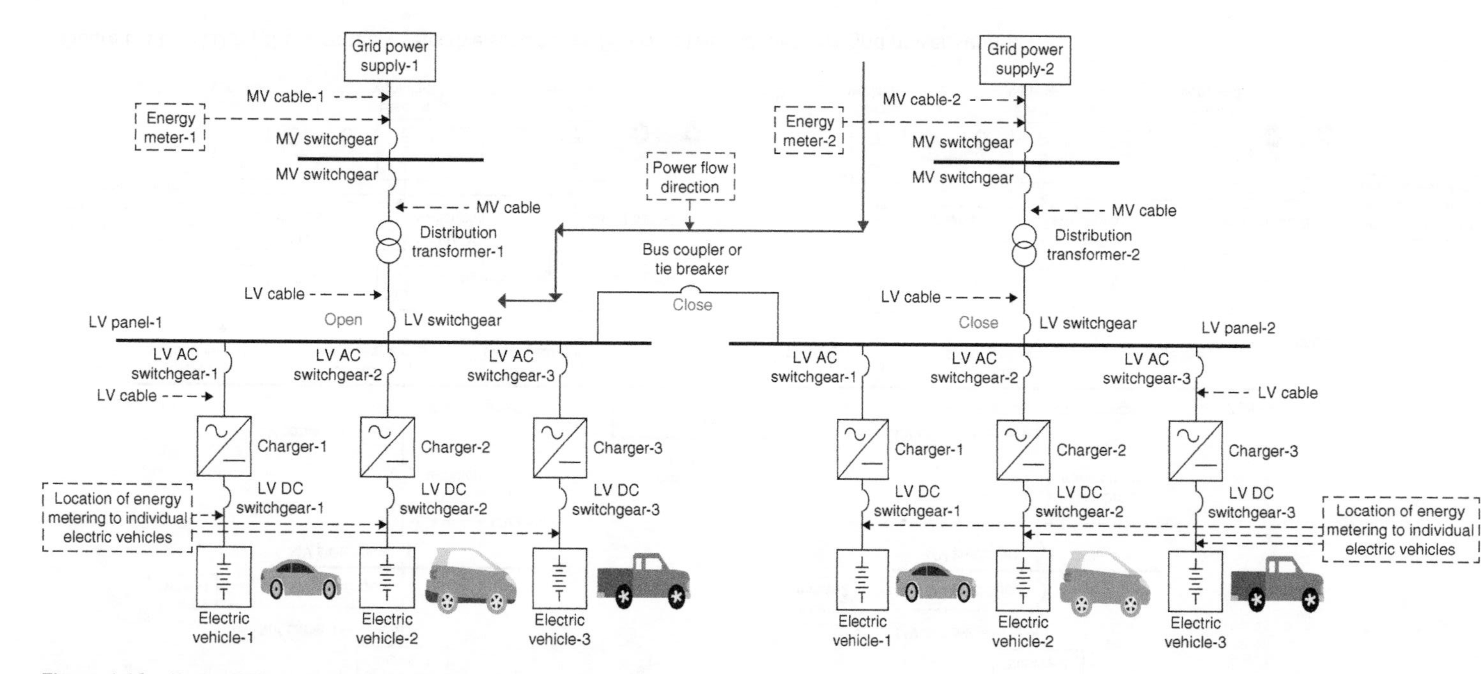

Figure 6.12 Entire FCS loads through distribution transformer-2.

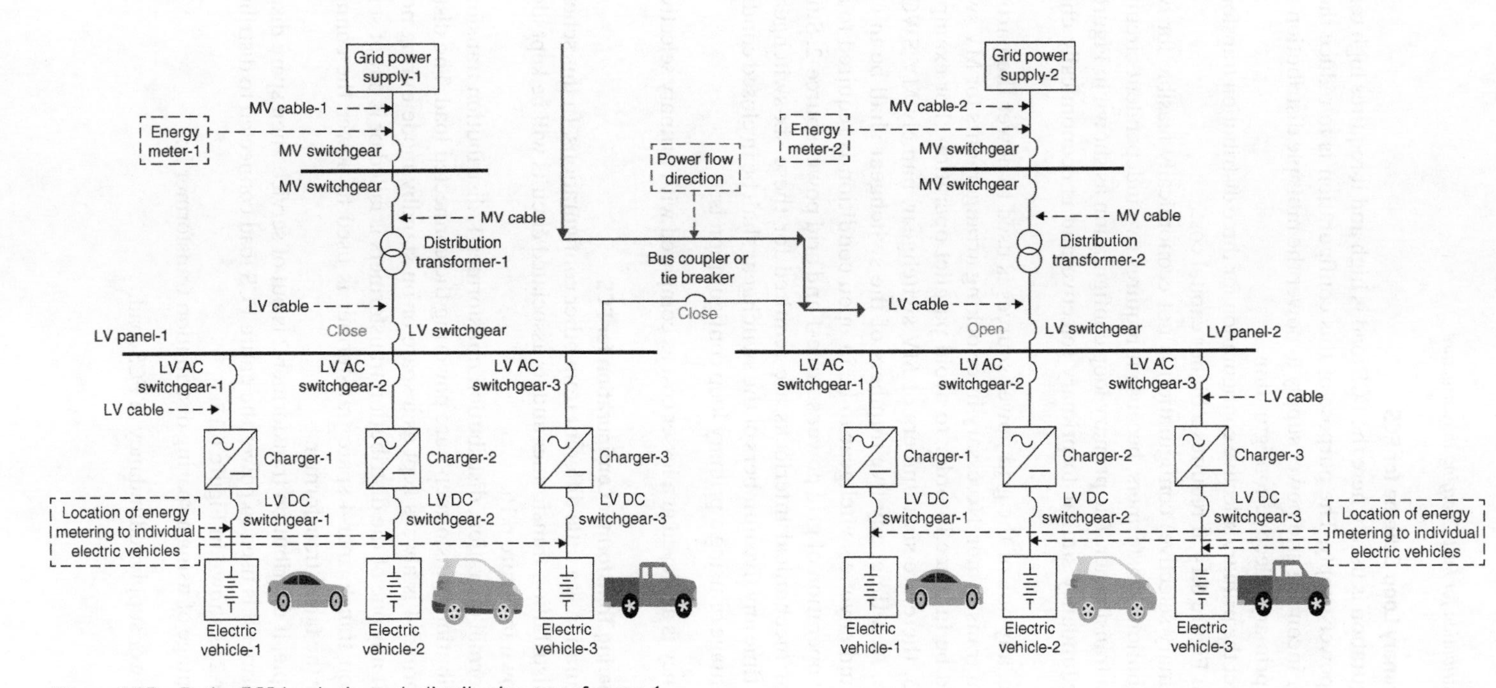

Figure 6.13 Entire FCS loads through distribution transformer-1.

6.3.2.5 Primary Loop Scheme for FCS

This configuration is used where the FCS load is high and it requires high reliability in grid power supply. The purpose of this configuration is to reduce the cost required for incoming grid power supply to power the multiple distribution transformers in primary selective configuration.

If we select the primary selective configuration for three distribution transformers as shown in Figure 6.14, it requires a higher capital cost.

This (primary selective) configuration is not economically feasible for two or more distribution transformers, because it requires two independent circuits per distribution transformer. The primary loop configuration as shown in Figure 6.15 provides redundancy similar to primary selective and is economically cheaper as well.

At a time, any one of the grid power sources is used to power the particular distribution transformer. Necessary interlocking arrangements for MV switchgear should be in place in order to avoid parallel operation. For example, in Figure 6.15, there are six numbers of MV switchgear, namely MV SWG-1 to MV SWG-6. At a time, any five numbers of the switchgear shall be in closed condition and anyone switchgear to be in open condition is required to avoid the parallel operation of grid power source-1 and grid power source-2. Suitable electrical or mechanical interlocks are required for these six switchgears to ensure at a time any five numbers of the switchgear shall be in closed condition.

The advantage of using a primary loop configuration is:

- Redundancy is achieved at a lesser cost as compared with primary selective.

6.3.2.6 Sparing Transformer Configuration for FCS

This configuration is a little different from other configurations. In this scheme, a dedicated distribution transformer and its associated circuits will be kept ideal as spare as shown in Figure 6.16.

During normal operation, distribution transformer-1, distribution transformer-2, and distribution transformer-3 are powering the connected load. The distribution transformer-4 is always kept as a spare or on standby mode during normal operation. If any one of the distribution transformers is failed or is out of service, a distribution transformer-4 spare transformer is used to power the connected loads under the failed transformer.

For example, if distribution transformer-2 is out of service, then spare distribution transformer-4 is used to power the entire FCS load connected to distribution transformer-2 as shown in Figure 6.17.

The advantage of using a sparing distribution transformer is:

- Higher power supply redundancy to FCS load.

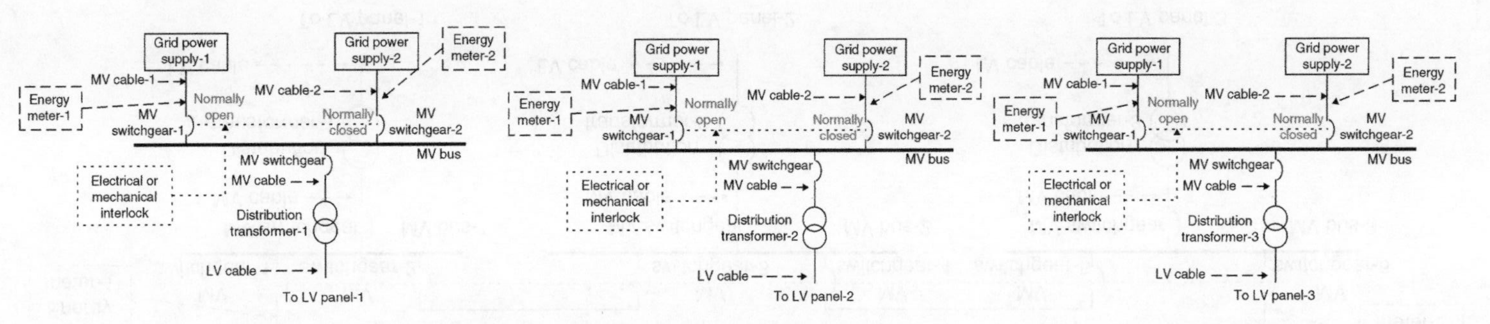

Figure 6.14 Primary selective configuration for three distribution transformers.

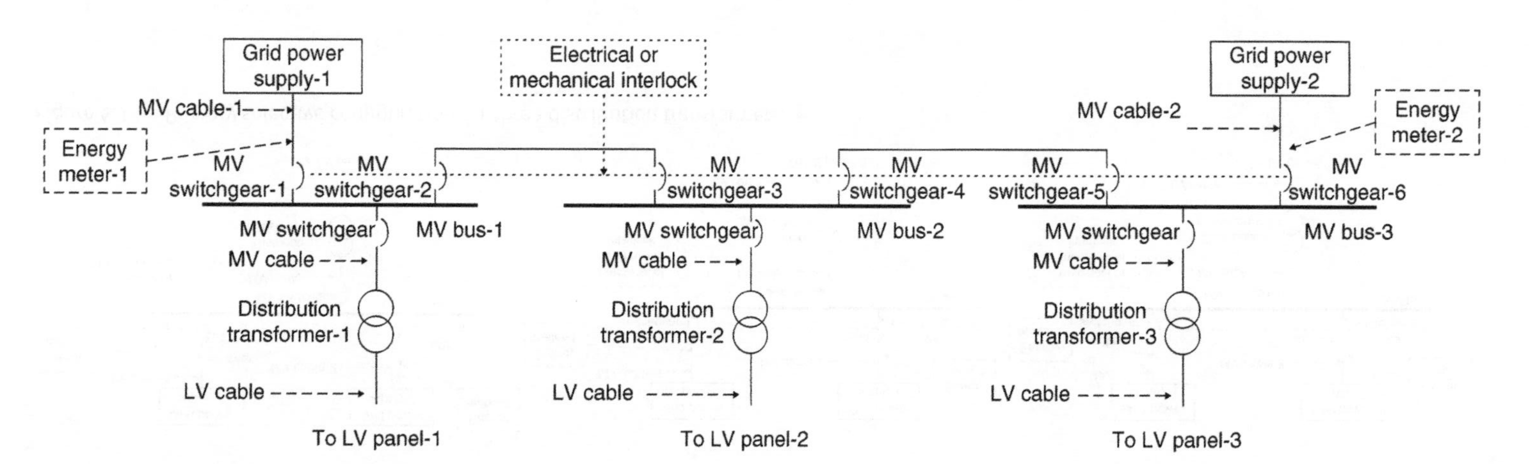

Figure 6.15 Primary loop configuration for three distribution transformers.

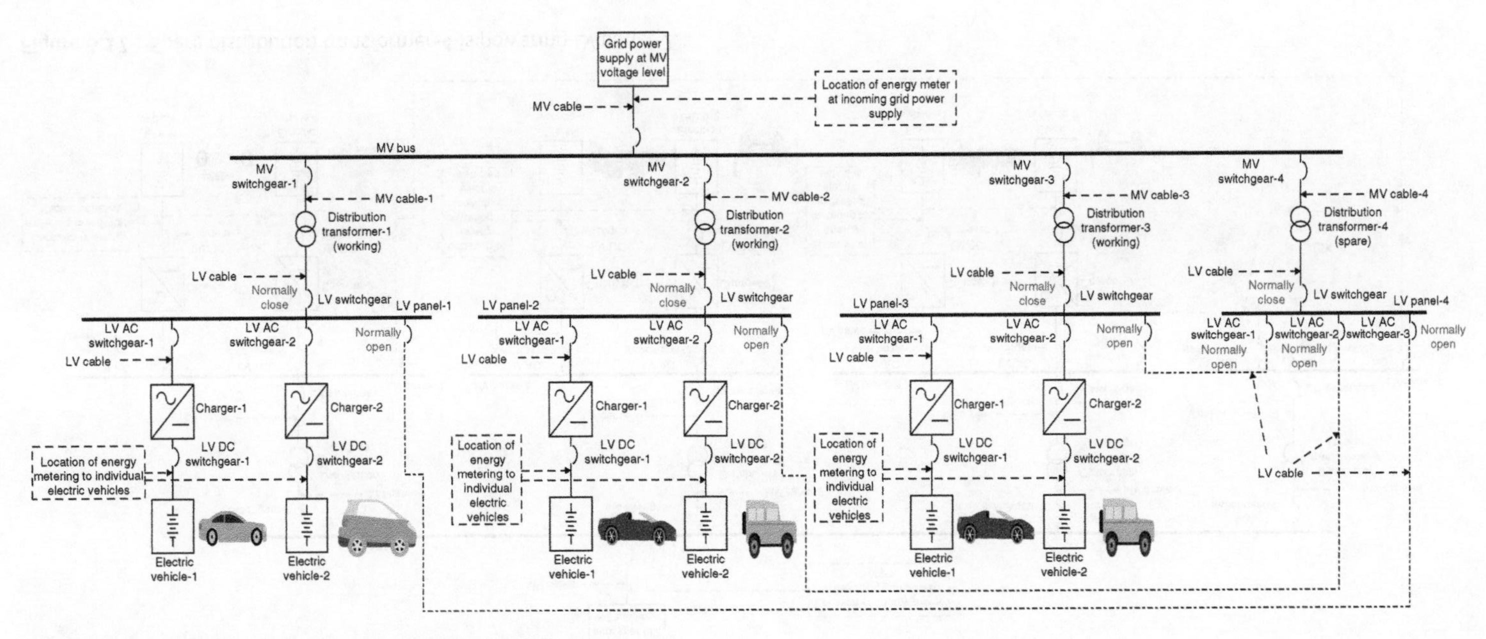

Figure 6.16 Sparing transformer configuration for FCS.

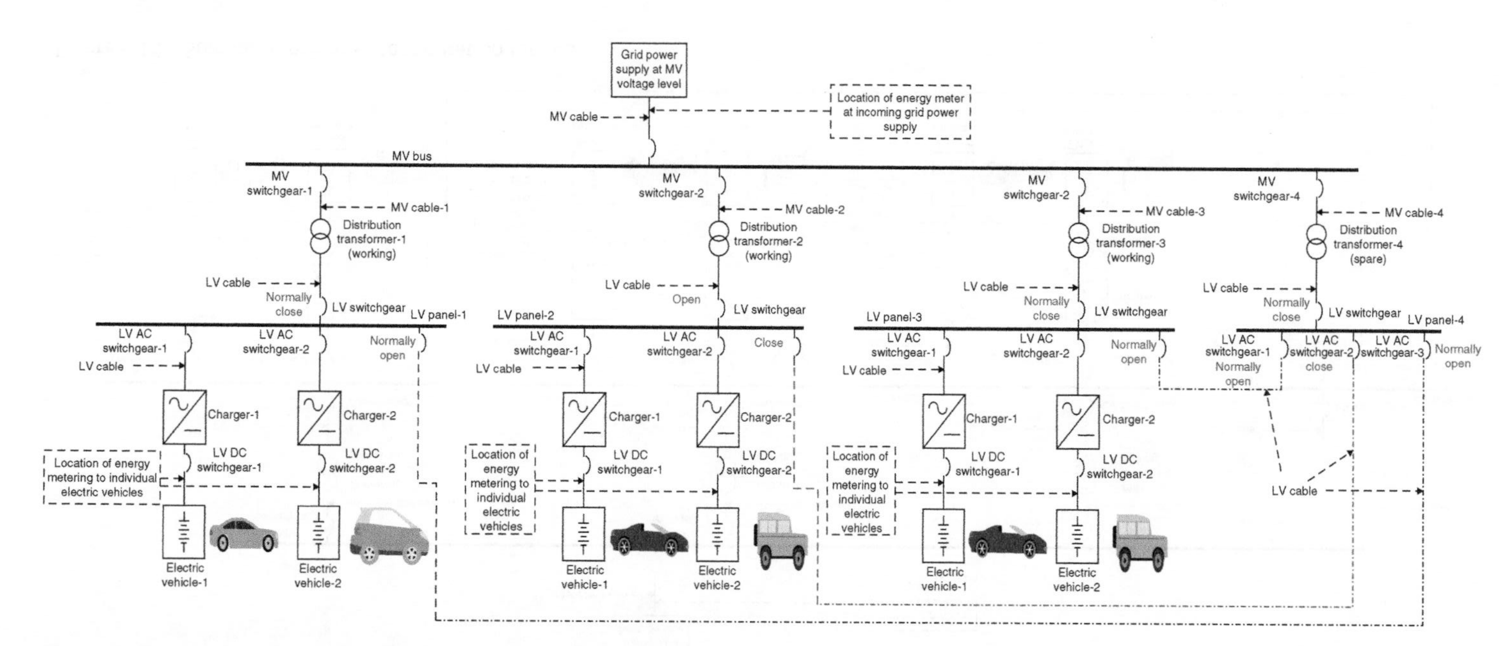

Figure 6.17 Spare distribution transformer-4 is powering LV panel-2.

The disadvantages of using a sparing distribution transformer scheme are:

- ❖ A dedicated distribution transformer to be kept spare and energized throughout the service. So, it will draw no-load loss from the supply throughout the service.
- ❖ Cost of spare distribution transformer and its circuit is additional.

6.3.2.7 Other Configurations for FCS

The typical radial power distribution from the LV panel to the charger is shown in Figure 6.18.

Failure of the distribution circuit to the charger (i.e. switchgear and cable) results in a power outage.

If the FCS requires higher reliability in the distribution circuit to the charger, then any of the following redundancy configurations can be used:

a) Redundancy through automatic transfer switch (ATS)
b) Redundancy through static transfer switch (STS)

Redundancy Through ATS

The typical redundancy configurations (configuration 1 and configuration 2) with ATS are shown in Figures 6.19 and 6.20.

Congratulation 1:

An ATS is electrical equipment used to select the power source which is used to power the FCS load. It has two incoming supply connections and one output connection. The incoming power source and load (charger) shall be connected as shown in Figure 6.19.

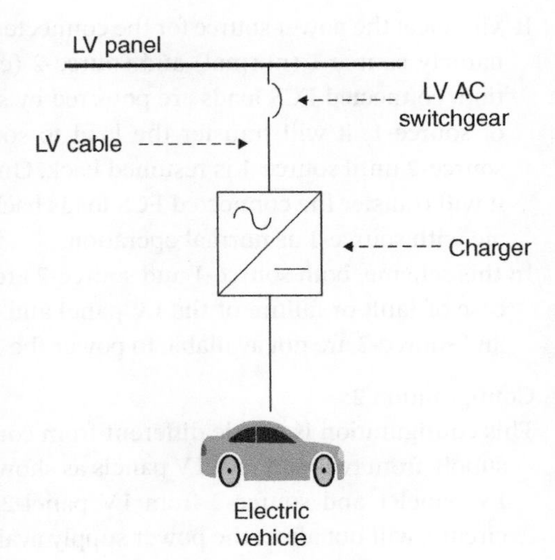

Figure 6.18 Typical radial power distribution from LV panel to charger.

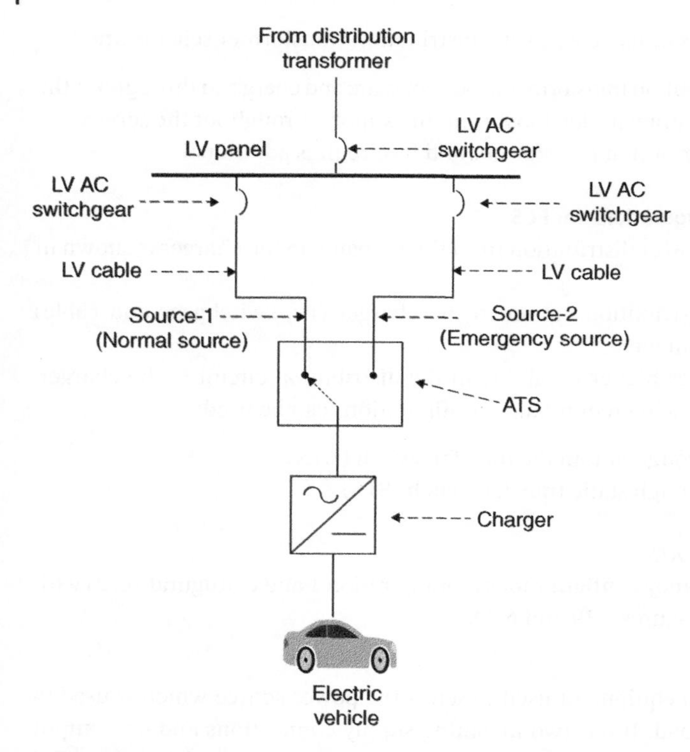

Figure 6.19 ATS for redundancy – configuration 1.

It will select the power source for the connected load. It has two incoming sources, namely source-1 (normal) and source-2 (emergency). During normal operation, connected FCS loads are powered by source-1. In the event of the failure of source-1, it will transfer the load to source-2 and continue to operate in source-2 until source-1 is resumed back. Once source-1 is resumed back, again it will transfer the connected FCS loads back to source-1 and continue to operate with source-1 as normal operation.

In this scheme, both source-1 and source-2 are taken from the same LV panel. In case of fault or failure of the LV panel and its incoming supply, both source-1 and source-2 are not available to power the FCS loads.

Configuration 2:

This configuration is a little different from configuration 1. It receives the power supply from two different LV panels as shown in Figure 6.20, i.e. source-1 from LV panel-1 and source-2 from LV panel-2. Failure of any LV panel and its circuits will not affect the power supply availability to FCS loads.

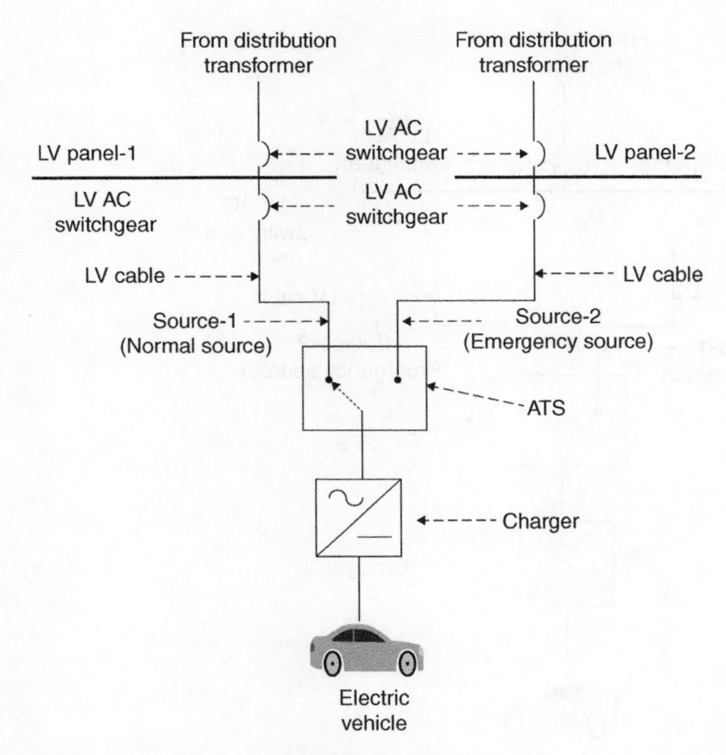

Figure 6.20 ATS for redundancy – configuration 2.

Redundancy Through STS

The difference between ATS and STS are:

- ATS uses the mechanical elements and STS uses the power semiconductor devices like silicon controlled rectifiers for transferring the load between sources. ATS will transfer the connected loads from source-1 to source-2 and vice versa with a break (in a few cycles to a second) in a open transition.
- STS will transfer the connected loads from source-1 to source-2 and vice versa without any break in a closed transition, i.e. it will transfer the load with sync.

In order to use the STS for circuit redundancy, the following conditions are always met for seamless transfer:

- The voltage of both source-1 and source-2 shall be the same.
- The frequency of both source-1 and source-2 shall be the same.
- The phase sequence of both source-1 and source-2 shall be the same.
- The phase angle of both source-1 and source-2 shall be the same.

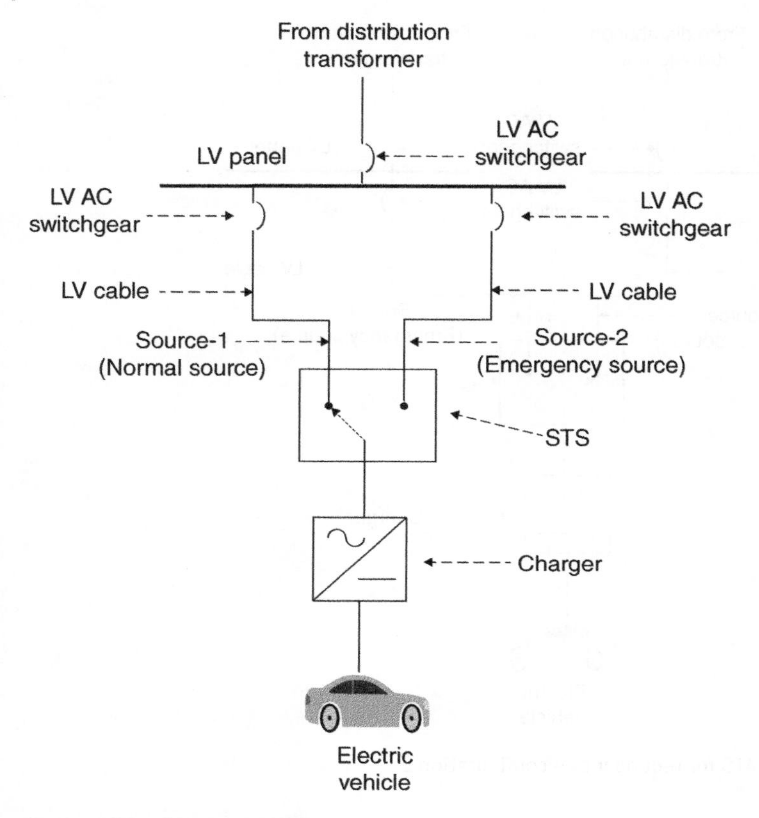

Figure 6.21 STS for redundancy – configuration 1.

The typical redundancy configurations (configuration 1 and configuration 2) with STS are shown in Figures 6.21 and 6.22.

Configuration 1:

A STS is an electrical component that is used to select the power source which would power the FCS load. It has two incoming supply connections and one output connection. The incoming power source and load (charger) shall be connected as shown in Figure 6.21. It is important to note that, in order to transfer the loads from source-1 to source-2 and vice versa, it requires meeting the synchronous conditions such as voltage, frequency, phase sequence, and phase angle are always the same. Since, both sources are extended from the same LV panel, except for phase sequence, other three parameters are almost same and synchronization between sources are generally not a problem.

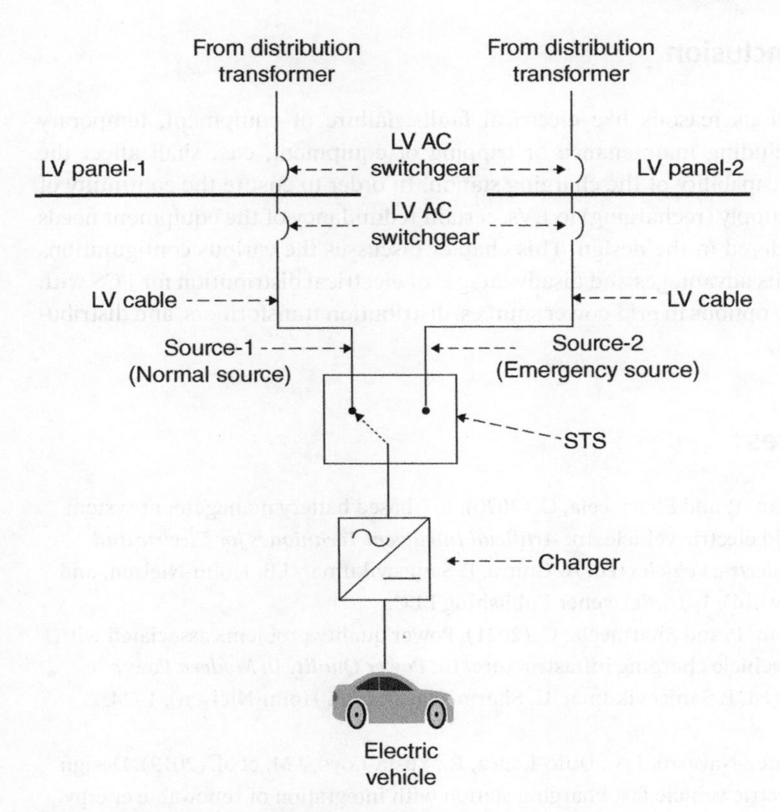

Figure 6.22 STS for redundancy – configuration 2.

In this configuration, both sources are extended from the same LV panel. When source-1 fails, STS will transfer the loads from source-1 to source-2 without any break and continue to operate with source-2 until source-1 is resumed back. Once, source-1 is resumed back, again STS will transfer back to source-1. The end users can select the priority source for their normal operation either source-1 or source-2.

Configuration 2:

In this configuration, STS will receive the power supply from two different LV panels, i.e. source-1 from LV panel-1 and source-2 from LV panel-2. It is important to note that, in order to transfer the loads from source-1 to source-2 and vice versa, it requires meeting the synchronous conditions such as voltage, frequency, phase sequence, and phase angle to always be the same. Hence, the conditions for synchronization are taken into consideration for designing the electrical distribution.

6.4 Conclusion

Due to various reasons like electrical faults, failure of equipment, temporary outages including maintenance or tripping of equipment, etc., shall affect the recharging capability of the charging station. In order to ensure the continuity of the power supply (recharging) to EVs, certain redundancy of the equipment needs to be considered in the design. This chapter discusses the various configurations along with its advantages and disadvantages of electrical distribution for FCS with redundancy options in grid power sources, distribution transformers, and distribution circuits.

References

1 Sivaraman, P. and Sharmeela, C. (2020). IoT-based battery management system for hybrid electric vehicle. In: *Artificial Intelligent Techniques for Electric and Hybrid Electric Vehicles* (ed. A. Chitra, P. Sanjeevikumar, J.B. Holm-Nielsen, and S. Himavathi), 1–16. Scrivener Publishing LLC.

2 Sivaraman, P. and Sharmeela, C. (2021). Power quality problems associated with electric vehicle charging infrastructure. In: *Power Quality in Modern Power Systems* (ed. P. Sanjeevikumar, C. Sharmeela, and J.B. Holm-Nielsen), 1–24. Elsevier.

3 Dominguez-Navarro, J.A., Dufo-Lopez, R., Yusta-Loyo, J.M. et al. (2019). Design of an electric vehicle fast-charging station with integration of renewable energy and energy storage systems. *Electrical Power and Energy Systems* 105: 46–58.

4 Bayram, I. and Tajer, A. (2017). *Plug-in Electric Vehicle Grid Integration*, 1e. Artech.

5 Gnann, T., Funke, S., Jakobson, N. et al. (2018). Fast charging infrastructure for electric vehicles: today's situation and future needs. *Transportation Research Part D* 62: 314–329.

6 Chowdhury, A. and Koval, D. (2009). *Power Distribution System Reliability: Practical Methods and Applications*, 1e. Wiley-IEEE Press.

7 Sivaraman, P. and Sharmeela, C. (2020). Existing issues associated with electric distribution system. In: *Handbook of Research on New Solutions and Technologies in Electrical Distribution Networks* (ed. B. Kahn, H.H. Alhelou, and G. Hayek), 1–31. Hershey, PA: IGI Global.

8 Singh, C., Jirutitijaroen, P., and Mitra, J. (2019). *Electric Power Grid Reliability Evaluation: Models and Methods*, 1e. Wiley-IEEE Press.

9 Daza, S.A. (2016). *Electric Power System Fundamentals*, 1e. Artech.

10 El-Hawary, M.E. (2008). *Introduction to Electrical Power Systems*, 1e. Wiley-IEEE Press.

11 Pansini, A.J. (2005). *Power Transmission and Distribution*, 1e. River Publishers.

12 Khan, S. (2008). *Industrial Power Systems*, 1e. CRC Press.

13 Billinton, R., Nerode, R., and Wood, A.J. (2003). *Power-System Reliability Calculations*, 1e. MIT Press.

14 IEEE Std 141-1993 (R1999) (1993). *IEEE Recommended Practice for Electric Power Distribution for Industrial Plants*. IEEE Standards

15 Sivaraman, P. and Sharmeela, C. (2020). Introduction to electric distribution system. In: *Handbook of Research on New Solutions and Technologies in Electrical Distribution Networks* (ed. B. Kahn, H.H. Alhelou, and G. Hayek), 1–31. Hershey, PA: IGI Global.

16 IEEE Std 241-1990 (1990). *IEEE Recommended Practice for Electric Power Systems in Commercial Buildings*. IEEE Standards

7

Energy Storage System for Fast-Charging Stations

7.1 Introduction

The fast-charging stations are widely used for recharging the EVs in lesser time duration (typically 30–60 minutes from 0% SoC to 100% SoC) [1]. These fast-charging stations are generally installed in public places like hotels, shopping malls, cinema theaters, hospitals, etc., and on highway roads to enable EV users to reach their destinations. In this method, EV batteries are charged with fast chargers which draw high power from the source and charge the EV batteries in a lesser time duration. The rating of fast chargers varies among different manufacturers and the typical power rating of fast chargers is in the range of 50–300 kW. Chargers will draw the power up to its rated capacity from the source for recharging the batteries during the charging period. These EV loads are totally new to the distribution system and the existing distribution system is not designed for this. Also, sometimes higher penetration of EV charging loads causes power imbalance (generation ≠ demand) in the power system. One more important point is that EV charging loads can be connected to any place or anywhere across the distribution system and it requires certain modifications like distribution transformer upgradations, distribution feeder upgradation, etc., to cater to the additional EV charging loads [2]. Distributed renewable energy sources (DRES) like solar PV systems are used to generate electric power locally closer to the loads to reduce losses in the distribution system and avoid issues related to voltage regulation in the distribution system [3, 4]. Also, it reduces the capacity expansion of the distribution system. Nowadays, fast-charging stations are using the DRES like rooftop solar PV systems to meet energy requirements. The problem with solar PV systems is that "output power is not constant throughout the operation" and is "not available

Fast-Charging Infrastructure for Electric and Hybrid Electric Vehicles: Methods for Large-Scale Penetration into Electric Distribution Networks, First Edition. Sivaraman Palanisamy, Sharmeela Chenniappan, and Sanjeevikumar Padmanaban.

Published 2023 by John Wiley & Sons, Inc.

during a night-time" [5, 14]. Also, the connected EV loads at any given point in time cannot be held constant.

In order to increase the charging station availability over an extended duration, ESS can be considered as one of the options to overcome the issues with solar PV systems. ESS plays an important role in smoothing the output power from the solar PV system and thereby maintaining the load-generation balance [6]. The ESS employed along with solar PV systems is used to store the energy whenever the generation from solar PV systems is higher than the fast-charging station load demand and energy stored in the ESS is used to charge the EVs whenever the power generation from the solar PV system is lesser than fast-charging station load demand.

The advantages of using the ESS in fast-charging stations are [7]:

- Providing the backup power
- Islanding mode of operation
- Capacity expansion deferral
- Peak load demand shaving
- Demand response
- Reduce the Peak–Valley difference
- Spinning reserve
- Smoothing the output of the renewable source
- Volt/Var support

This chapter discusses the various benefits and operating scenarios that can be achieved by having the energy storage system employed along with renewable energy sources, microgrids, and distribution systems. ESS, in general, helps to enhance the performance, reliability, availability, and operation of fast-charging stations.

7.2 Renewables + ESS

The EVs are alternatives to conventional petrol and diesel vehicles. These EVs will not produce any carbon emissions during their operation. But, if the EVs are recharged with electric power generated from conventional coal- or diesel-based power plants, if the EVs entire ecosystem is charged using electricity from conventional generation, then the whole idea of a green future remains defeated [8]. The carbon emission associated with EV charging in the ecosystem can be reduced by means of using power from renewable energy sources. The output power from renewable energy sources is clean and it does not involve any carbon emission during its operation. Renewable energy sources like solar PV systems and wind energy conversion are widely used for power generation.

The chapter further discusses the different schemes, of using ESS with renewables which enhances the flexibility of operation in the overall schemes, and also showcases how ESS can help reduce the intermittent nature of the renewable source availability. This in turn enhances the reliability of the fast-charging station.

7.2.1 Solar PV System without Battery Energy Storage System – Scheme 1 AC Interconnection

The typical single line diagram (SLD) of fast-charging station loads is powered through a solar PV system – scheme 1 AC interconnection is shown in Figure 7.1. The fast-charging station receives the utility power supply from DISCOM at medium voltage (MV) and it is stepped down to the low voltage (LV) by means of a step-down distribution transformer. The solar PV systems are connected to the LV bus (AC) of the fast-charging station.

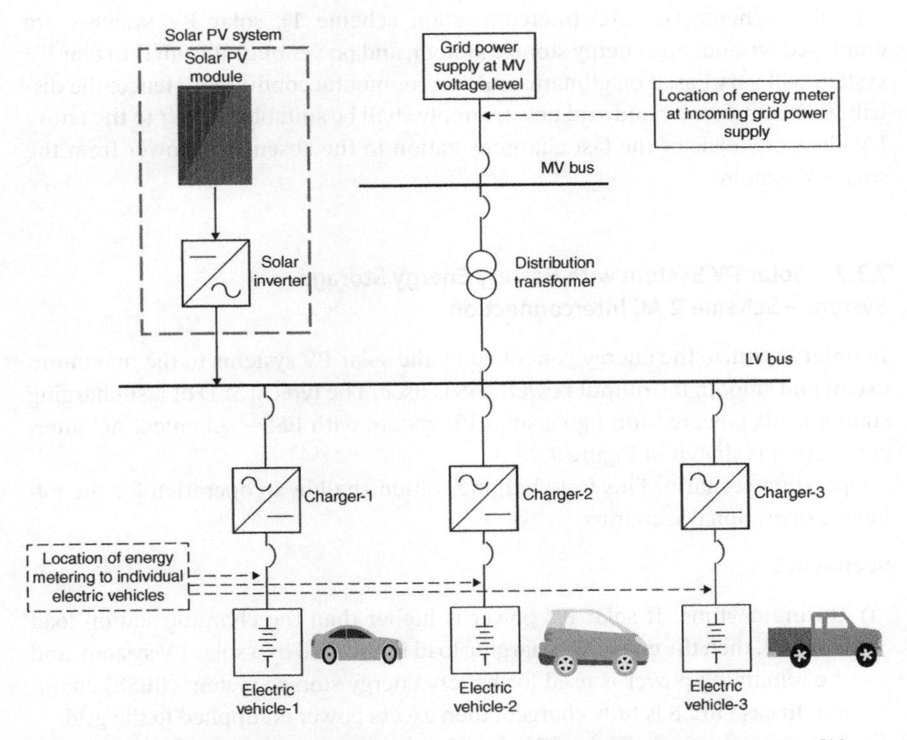

Figure 7.1 Typical SLD of fast-charging station loads is powered through a solar PV system – scheme 1 AC interconnection.

Operation scenario: This fast-charging station shall be in operation for the following operational scenarios:

Scenario 1

i) During daytime: Entire EV charging load powered by the solar PV system.
ii) During night-time: Entire EV charging loads are powered by grid.

Scenario 2

i) During daytime: Partial EV charging load powered by solar PV system + partial EV charging loads are powered by grid.
ii) During night-time: Entire EV charging loads are powered by grid.

Scenario 3

i) During daytime: In the absence of power from a solar PV system, the entire EV charging load is powered by the grid.
ii) During night-time: Entire EV charging loads are powered by grid.

In this scheme (i.e. AC interconnection scheme 1), solar PV systems are employed without any energy storage system, and power output from the solar PV system will vary based on climatic and environmental conditions. Hence, the distribution transformer and grid power supply shall be suitable to cater to the entire EV charging loads of the fast-charging station in the absence of power from the solar PV system.

7.2.2 Solar PV System with Battery Energy Storage System – Scheme 2 AC Interconnection

In order to utilize the energy generated by the solar PV systems to the maximum extent and smooth the output power, ESS is used. The typical SLD of fast-charging station loads powered through a solar PV system with ESS – scheme 2 AC interconnection is shown in Figure 7.2.

Operation scenario: This fast-charging station shall be in operation for the following operational scenarios:

Scenario 1

i) During daytime: If solar PV power is higher than the charging station load demand, then the entire EV charging load is powered by a solar PV system, and the remaining power is used for battery energy storage system (BESS) charging. In case BESS is fully charged, then excess power is supplied to the grid.
ii) During night-time: Entire EV charging loads are powered by grid + BESS discharging or only grid in case BESS is completely discharged.

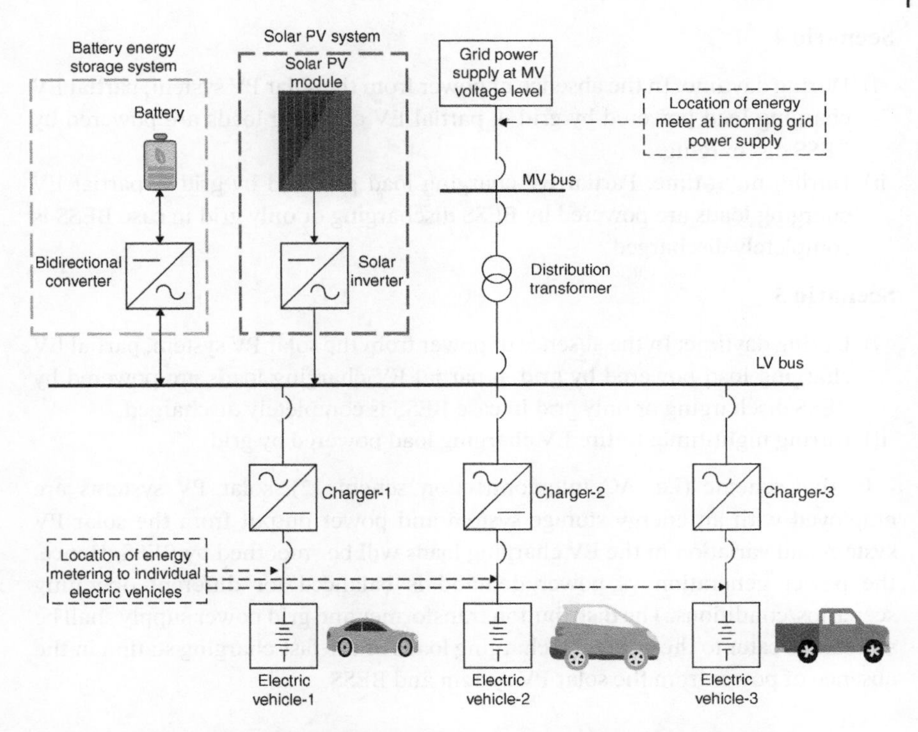

Figure 7.2 Typical SLD of fast-charging station loads is powered through a solar PV system with ESS – scheme 2 AC interconnection.

Scenario 2

i) During daytime: If solar PV power is equal to the charging station load demand, then the entire EV charging load is powered by the solar PV system.
ii) During night-time: EV charging loads are powered by grid + BESS or only grid in case BESS is completely discharged.

Scenario 3

i) During daytime: If solar PV power is lesser than charging station load demand, partial EV charging load powered by solar PV system + partial EV charging loads are powered by BESS discharging + partial EV charging loads are powered by grid.
ii) During night-time: Partial EV charging loads are powered by grid + partial EV charging loads are powered by BESS discharging or only grid in case BESS is completely discharged.

Scenario 4

i) During daytime: In the absence of power from the solar PV system, partial EV charging load powered by grid + partial EV charging loads are powered by BESS discharging.
ii) During night-time: Partial EV charging load powered by grid + partial EV charging loads are powered by BESS discharging or only grid in case BESS is completely discharged.

Scenario 5

i) During daytime: In the absence of power from the solar PV system, partial EV charging load powered by grid + partial EV charging loads are powered by BESS discharging or only grid in case BESS is completely discharged.
ii) During night-time: Entire EV charging load powered by grid.

In this scheme (i.e. AC interconnection scheme 2), solar PV systems are employed with an energy storage system and power output from the solar PV system and variation in the EV charging loads will be smoothed by BESS. Hence, the power generation vs power demand is balanced for different operating scenarios/conditions. The distribution transformer and grid power supply shall be suitable to cater to the entire EV charging loads of the fast-charging station in the absence of power from the solar PV system and BESS.

7.2.3 Solar PV System with Battery Energy Storage System – Scheme 3 DC Interconnection

Alternative configuration to scheme 1 and scheme 2 is shown in Figure 7.3 as scheme 3 DC interconnection. The nature of EV charging is DC electricity and the nature of solar PV generation and BESS is also DC electricity, hence DC interconnection is employed for solar PV systems and BESS in the fast-charging station. The typical SLD of fast-charging station loads with DC interconnection is shown in Figure 7.3.

Operation scenario: This fast-charging station shall be in operation for the following operational scenarios:

Scenario 1

i) During daytime: If solar PV power is higher than the charging station load demand, then the entire EV charging load is powered by solar PV system + BESS charging.
ii) During night-time: Entire EV charging loads are powered by grid + BESS discharging or only grid in case BESS is completely discharged.

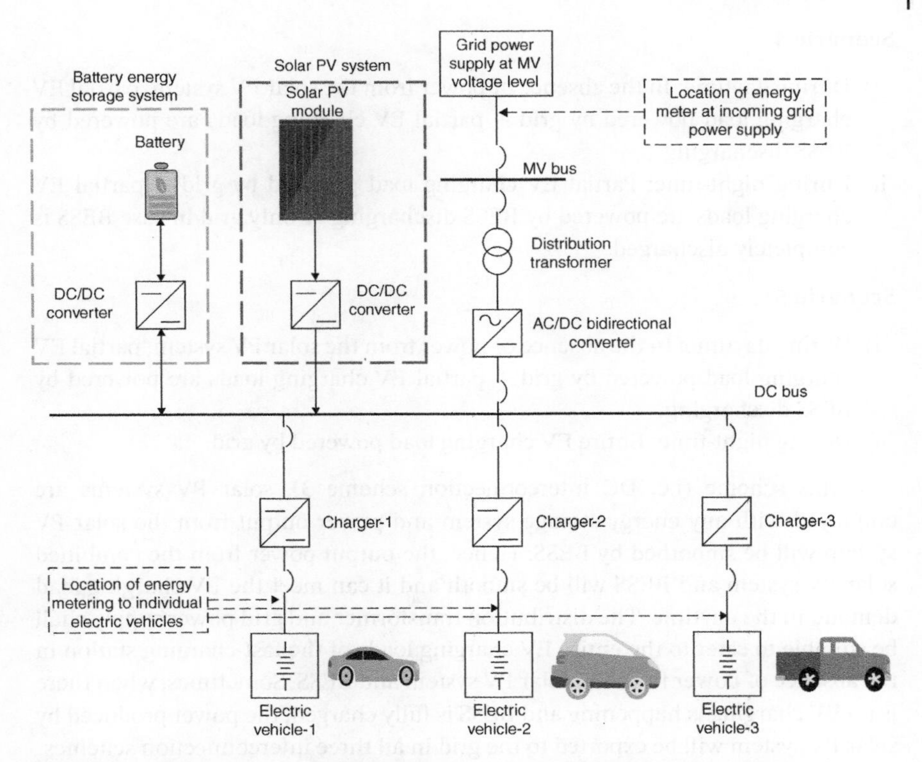

Figure 7.3 Typical SLD of fast-charging station loads is powered through a solar PV system and BESS – scheme 3 DC interconnection.

Scenario 2

i) During daytime: If solar PV power is equal to the charging station load demand, then the entire EV charging load is powered by the solar PV system.
ii) During night-time: Partial EV charging loads are powered by grid + partial EV charging loads are powered by BESS discharging or only grid in case BESS is completely discharged.

Scenario 3

i) During daytime: If solar PV power is lesser than charging station load demand, partial EV charging load powered by solar PV system + partial EV charging loads are powered by BESS discharging + partial EV charging loads are powered by grid.
ii) During night-time: Partial EV charging loads are powered by grid + partial EV charging loads are powered by BESS discharging or only grid in case BESS is completely discharged.

Scenario 4

i) During daytime: In the absence of power from the solar PV system, partial EV charging load powered by grid + partial EV charging loads are powered by BESS discharging.
ii) During night-time: Partial EV charging load powered by grid + partial EV charging loads are powered by BESS discharging or only grid in case BESS is completely discharged.

Scenario 5

i) During daytime: In the absence of power from the solar PV system, partial EV charging load powered by grid + partial EV charging loads are powered by BESS discharging.
ii) During night-time: Entire EV charging load powered by grid.

In this scheme (i.e. DC interconnection scheme 3), solar PV systems are employed with any energy storage system and power output from the solar PV system will be smoothed by BESS. Hence, the output power from the combined solar PV system and BESS will be smooth and it can meet the EV charging load demand in the daytime. The distribution transformer and grid power supply shall be suitable to cater to the entire EV charging loads of the fast-charging station in the absence of power from the solar PV system and BESS. Sometimes, when there is no EV charging is happening and BESS is fully charged, the power produced by solar PV system will be exported to the grid in all three interconnection schemes.

7.3 Microgrid with Renewables + ESS

The IEEE Std 2030.8-2018 defines the microgrid as "A group of interconnected loads and distributed energy resources (DER) with clearly defined electrical boundaries that act as a single controllable entity with respect to the grid and can connect and disconnect from the grid to enable it to operate in both grid-connected or islanded modes" [9]. A microgrid has the capability to operate and maintain its own voltage and frequency. The microgrids are classified into two types, namely grid-connected microgrids and standalone microgrids. In general, grid-connected microgrids can operate in grid-connected or in islanded mode while standalone microgrids will operate without grid connection, i.e. the system is completely islanded [10, 15]. DER and ESS play an important role in both grid-connected and standalone microgrids in terms of steady-state and dynamic performance. The main advantages of using ESS are maintaining the load-generation balance, frequency of the microgrid, power smoothing, load shifting, peak shaving, etc.

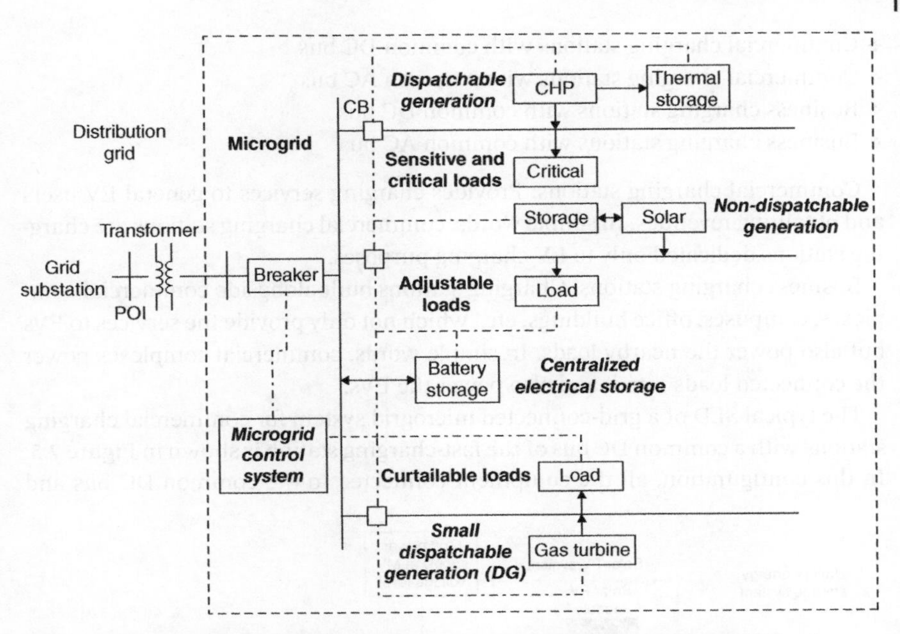

Figure 7.4 General structure of grid-connected microgrid [9] / with permission of IEEE. CHP, combined heat and power; POI, point of interconnection; CB, circuit breaker; DG, dispatchable generation.

The general structure of the grid-connected microgrid system as per IEEE Std 2030.8-2018 [9] is shown in Figure 7.4.

Both grid-connected and standalone microgrids are suitable for fast-charging stations to charge EVs. The selection of a grid-connected microgrid or standalone microgrid for the fast-charging station depends on multiple factors like availability of an electric grid nearby the vicinity or area, cost of electricity consumption from grid/kWH, the possibility of installation of DER locally, and DER installation capacity. In a standalone microgrid system, BESS is a key parameter that decides how many numbers of EVs can be charged at a time. On the other hand, a grid-connected microgrid system can be used for the rated number of EV charging even if the installed solar PV system and BESS are not sufficient to meet all EV loads. Generally, grid-connected microgrids are used for locations with high traffic flow like highways.

7.3.1 Grid-Connected Microgrid for Fast-Charging Stations

The IEC TR 62933-2-200:2021 brought the four typical application scenarios of grid-connected solar PV with an energy storage system powering the EV charging stations, namely [11]

- Commercial charging stations with common DC bus
- Commercial charging stations with common AC bus
- Business charging stations with common DC bus
- Business charging stations with common AC bus

Commercial charging stations: Provides charging services to general EV users and obtaining revenues. In simple words, commercial charging stations are charging stations dedicated only to EV charging purposes.

Business charging stations: Charging stations built alongside commercial complexes, campuses, office buildings, etc., which not only provide the services to EVs but also power the nearby loads. In simple words, commercial complexes power the connected loads into it and also power the EVs.

The typical SLD of a grid-connected microgrid system for commercial charging stations with a common DC bus of the fast-charging station is shown in Figure 7.5. In this configuration, all the equipment connected to the common DC bus and

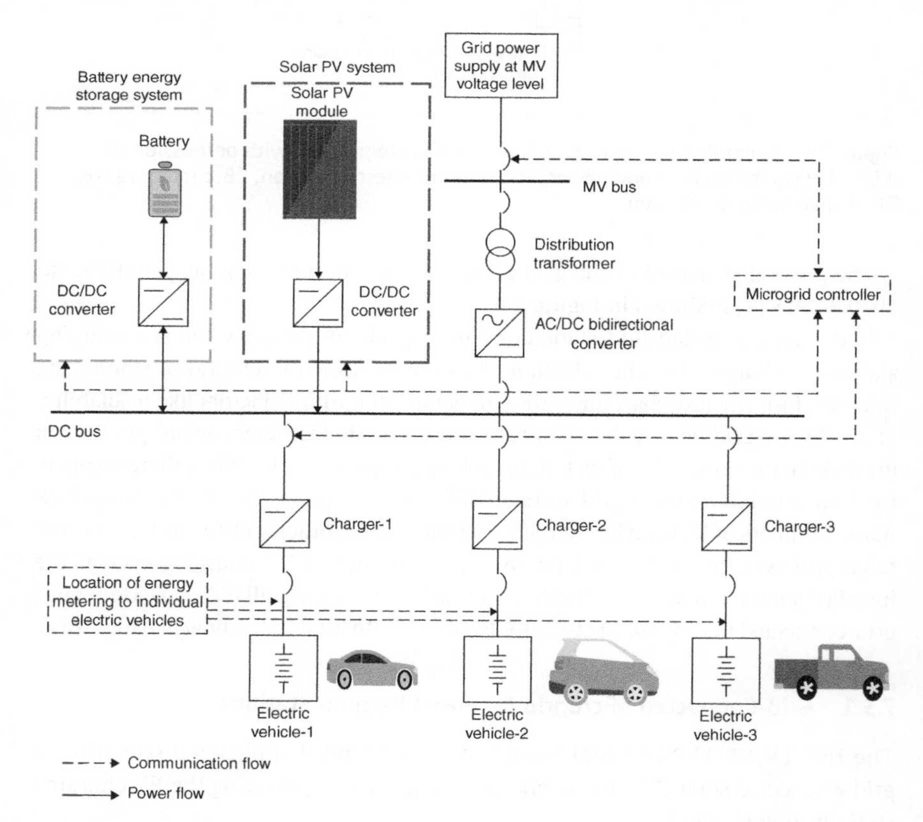

Figure 7.5 Typical SLD of grid-connected microgrid system for a commercial charging station with a common DC bus.

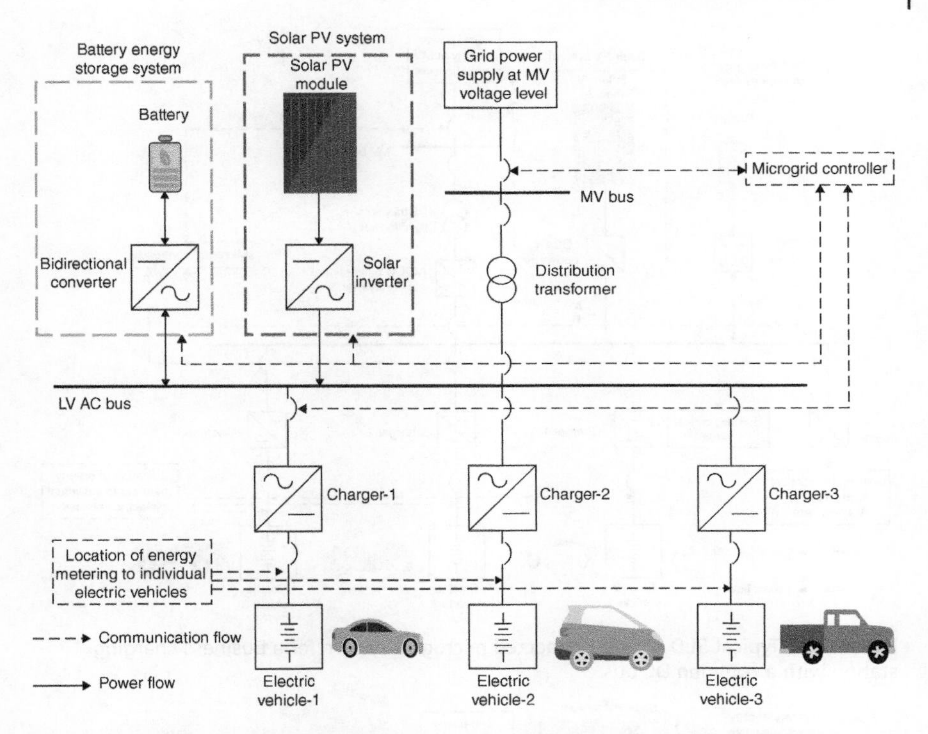

Figure 7.6 Typical SLD of grid-connected microgrid system for a commercial charging station with a common AC bus.

incoming grid power supply is converted into DC from AC by means of an AC–DC bidirectional converter as shown in Figure 7.5.

The typical SLD of a grid-connected microgrid system for commercial charging stations with a common AC bus of the fast-charging station is shown in Figure 7.6. In this configuration, all the equipment is connected to the common AC bus.

The typical SLD of a grid-connected microgrid system for business charging stations with a common DC bus of the fast-charging station is shown in Figure 7.7. In this configuration, all the equipment connected to the common DC bus and incoming grid power supply is converted into DC from AC by means of an AC–DC bidirectional converter as shown in Figure 7.7. If the building critical loads and noncritical loads operate in DC, then they will be directly connected to common DC without any additional converter for interface and if the building critical loads and noncritical loads operate in AC, then an additional DC–AC inverter interface is used to interconnect with common DC bus.

The typical SLD of a grid-connected microgrid system for business charging stations with a common AC bus of the fast-charging station is shown in Figure 7.8.

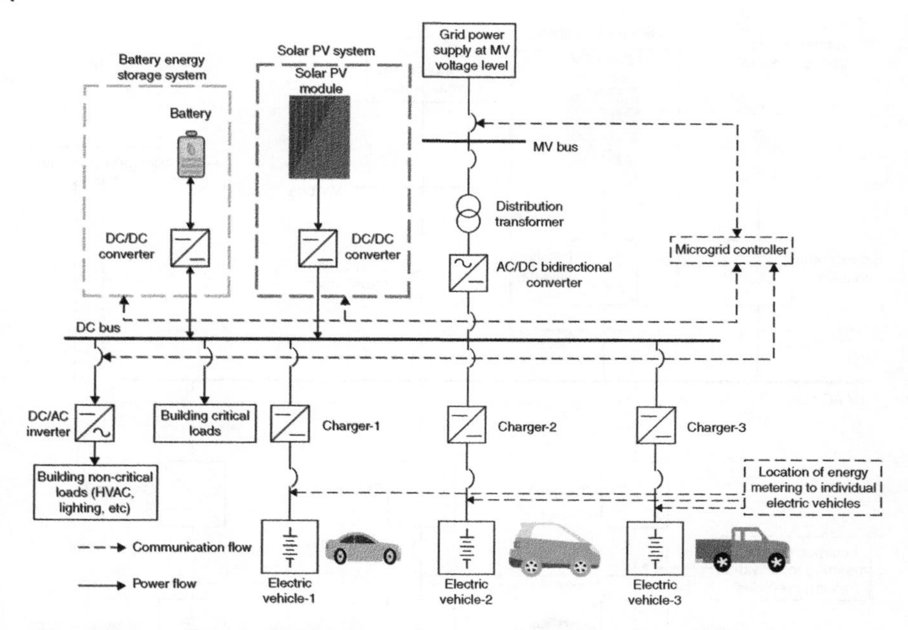

Figure 7.7 Typical SLD of grid-connected microgrid system for a business charging station with a common DC bus.

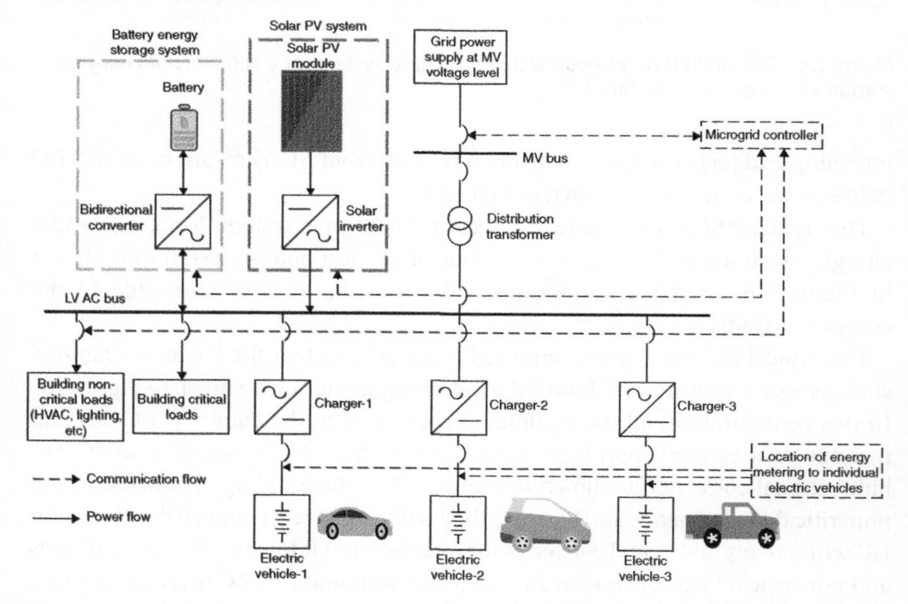

Figure 7.8 Typical SLD of grid-connected microgrid system for a business charging station with a common AC bus.

In this configuration, all the equipment connected to the common AC bus as shown in Figure 7.8. If the building critical loads and noncritical loads operate in AC, then they will be directly connected to common AC without any additional converter for interface and if the building critical loads and noncritical loads operate in DC, then an additional AC–DC rectifier interface is used to interconnect with common AC bus.

The grid-connected microgrid system for fast charging has the following elements:

- Solar PV system
- BESS
- Microgrid controller and its Communication interfaces
- Grid power supply
- Distribution transformer
- Switchgears
- Chargers

Solar PV systems are used to generate electricity from sunlight in and around the fast-charging stations. It could be either a rooftop solar PV system or a ground-mounted solar PV system or a combination of both. The type of solar PV system selected for the particular fast-charging station depends on total charging station loads and land/rooftop availability to install the solar PV module.

The BESS system is used to perform both charging and discharging functions to maintain the power balance throughout the operation, maximum utilization of solar energy, peak shaving, and ancillary services like voltage and frequency regulation. Generally, BESS will be in charging mode whenever solar PV power is higher than load demand during the daytime. It will be in discharge mode whenever the solar PV power is lesser than the load demand during daytime and nighttime to recharge the EVs. BESS overall serves the purpose of reducing the dependency on the grid.

The microgrid controller is the heart of the overall control system and it communicates with the solar PV system, BESS, and switchgear bidirectionally. It will receive the inputs like state of position (open/close) from switchgear, present active and reactive power generation from solar PV system and BESS, present load demand, etc., and it will take the actions like open/close command to switchgear, increase/decrease the active and reactive power flow, perform demand side management by managing the load demand as per energy availability, etc.

The grid power supply is used to power the fast-charging station loads whenever the solar PV + BESS is not able to meet the load demand. On the other hand, if solar PV power is higher than load demand and BESS is fully charged, then the remaining power from solar PV is exported to the grid. In simple words, the grid power supply is the bidirectional, it provides the power supply to the loads whenever the station needs it and whenever the solar PV system in the station produces more power than load demand + BESS charging, excess power will be exported back to the grid.

A distribution transformer is used for voltage transformation (i.e. step down the MV supply to the LV supply) to power all the connected loads [12].

The switchgear is used to open and close the circuit whenever needed by means of manual operation and open the circuit during any fault. Also, the switchgear is communicated with the microgrid controller bidirectionally, and it opens and closes the circuit based on the command from the microgrid controller.

Chargers are the main loads of fast-charging station, and it is used to charge the EV batteries.

7.3.2 Standalone Microgrid for Fast-Charging Stations

The standalone microgrids are used to power the fast-charging station loads where the grid power supply is not available nearby and/or the cost of establishing the grid infrastructure is high. These standalone microgrids have the capability to operate without any incoming grid power supply. In these stations, a solar PV system provides the initial power/energy and BESS is used to store it and used for charging the EVs in a timely manner. The maximum number of EVs charging at a time depends on the BESS capacity and its state of charge. This is the major limitation of using this configuration. Generally, to enhance power availability, it is required to size the BESS sufficiently large to compensate for the nonavailability of the grid.

The typical SLD of a standalone microgrid system for fast-charging stations with a common AC bus is shown in Figure 7.9.

The typical SLD of a standalone microgrid system for fast-charging stations with a common DC bus is shown in Figure 7.10.

7.4 ESS Modes of Operation

The IEEE Std 2836-2021 listed the typical three operating modes of ESS system for electric vehicle charging station with solar PV systems as follows [13]:

- Power smoothing
- Load shaping
- Time of Use (ToU) price

The power smoothing mode is used for smoothing the sudden load change of the fast-charging station and reducing the fast-charging station varying load impact on the distribution system.

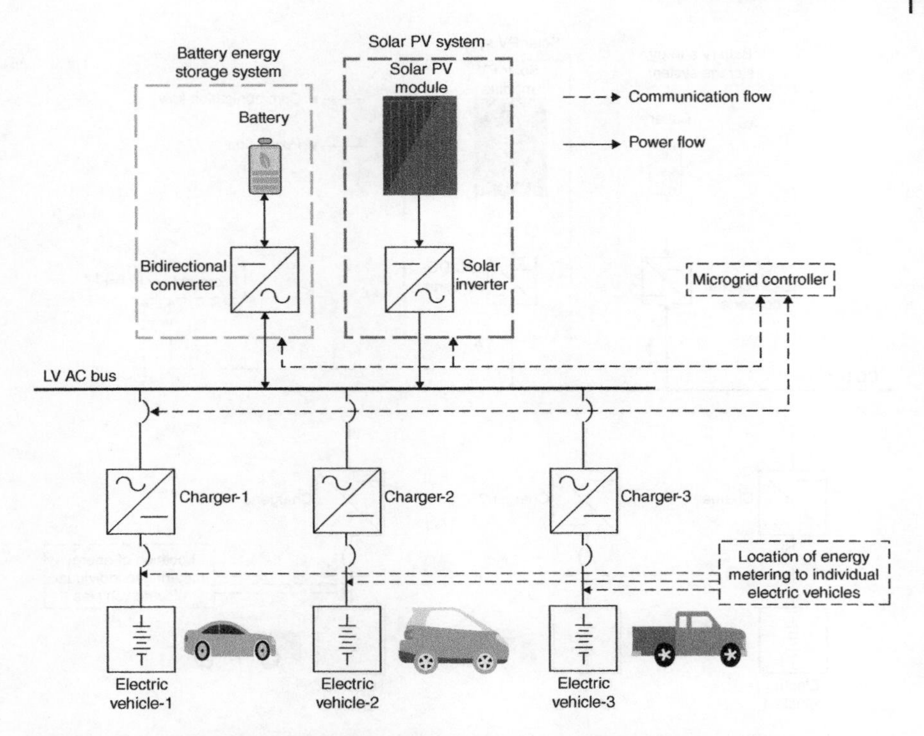

Figure 7.9 Typical SLD of standalone microgrid system with a common AC bus.

The load shaping mode is used for reducing the difference of power at peak time and off-peak time by absorbing the energy during off-peak time (i.e. charging at off-peak time) and discharging the energy during peak time. This mode reduces the peak load demand requirement of the fast-charging station and postpones the associated distribution infrastructure upgradation.

ToU price mode is used to get the benefits of energy price per kWh during its operation. The tariff structure of DISCOMs provides the different energy cost for peak hours and off-peak hours. During the peak hours, the energy cost/kWh is high and on the other hand, energy cost/kWh is low during off-peak hours. In order to get these benefits, ESS can be charged during off-peak hours at lesser cost and discharge it during peak hours at higher cost.

Duty cycle of an ESS in a fast-charging station is defined as the sequence of load an ESS is expected to supply for specified time periods. In simple words, duty cycle is a charge/discharge profile that is placed on an ESS. The general procedure to configure the duty cycle of an ESS for EV charging stations with solar PV system is shown in Figure 7.11 [13].

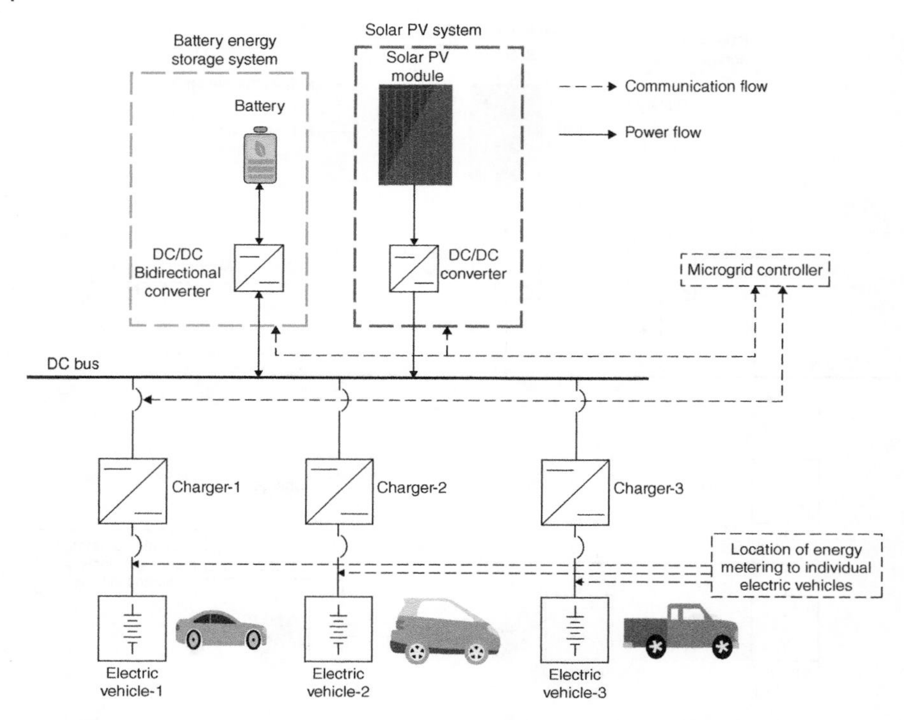

Figure 7.10 Typical SLD of standalone microgrid system with a common DC bus.

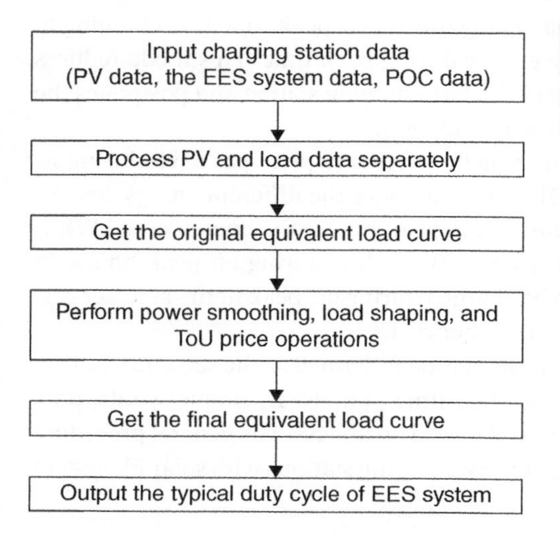

Figure 7.11 Procedure for configuring the duty cycle of an ESS [13] / with permission of IEEE. POC, point of connection.

Table 7.1 Example of duty cycle of an ESS in EV charging station.

S. No	Modes	Time	Target
1	Power smoothing	0:00–24:00	Smoothing power fluctuation of the load throughout the day
2	Load shaping	6:00–24:00	Reducing the peak–valley difference
3	ToU price	0:00–6:00	Charging during low price periods to gain benefits

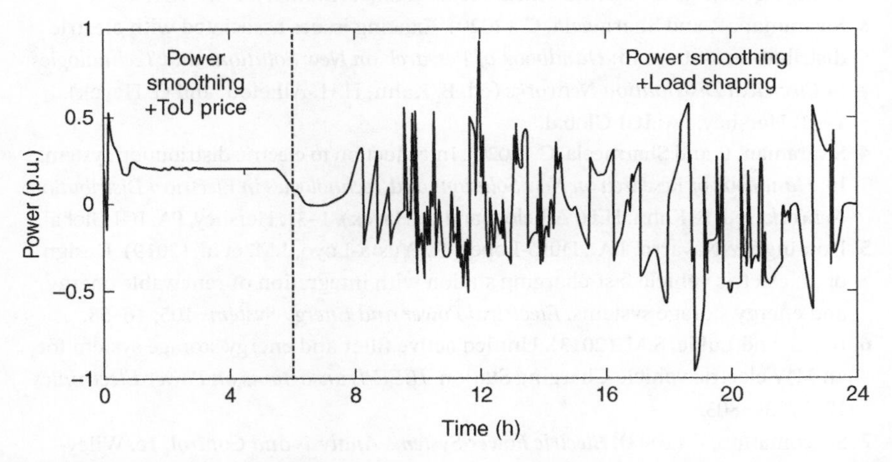

Figure 7.12 Typical duty cycle example curve of ESS in an EV charging station [13] / with permission of IEEE.

The typical example of time division of duty cycle considering three modes of operation for 24 hours is listed in Table 7.1.

An example for duty cycle curve for the ESS in an EV charging with solar PV system is shown in Figure 7.12 [13].

7.5 Conclusion

This chapter discusses the various AC and DC interconnection scheme of renewable energy sources and BESS for fast-charging station and highlights the flexibility in operation through various operating scenarios for each of the schemes. Also, the microgrid infrastructure with renewable energy sources and BESS for grid-connected and standalone microgrids is showcased. The benefits of ESS in fast-charging station and its advantage through different modes of operation such as peak shaving, load shaping, and ToU price are explained in detail.

References

1 Sivaraman, P. and Sharmeela, C. (2021). Power quality problems associated with electric vehicle charging infrastructure. In: *Power Quality in Modern Power Systems* (ed. P. Sanjeevikumar, C. Sharmeela, and J.B. Holm-Nielsen). London: Academic Press. https://doi.org/10.1016/B978-0-12-823346-7.00005-0.

2 Sivaraman, P., Sharmeela, C., and Logeshkumar, S. (2021). Charging infrastructure layout and planning for plug-in electric vehicles. In: *Cable Based and Wireless Charging Systems for Electric Vehicles* (ed. P. Sanjeevikumar), 1–24. IET.

3 Sivaraman, P. and Sharmeela, C. (2020). Existing issues associated with electric distribution system. In: *Handbook of Research on New Solutions and Technologies in Electrical Distribution Networks* (ed. B. Kahn, H.H. Alhelou, and G. Hayek), 1–31. Hershey, PA: IGI Global.

4 Sivaraman, P. and Sharmeela, C. (2020). Introduction to electric distribution system. In: *Handbook of Research on New Solutions and Technologies in Electrical Distribution Networks* (ed. B. Kahn, H.H. Alhelou, and G. Hayek), 1–31. Hershey, PA: IGI Global.

5 Dominguez-Navarro, J.A., Dufo-Lopez, R., Yusta-Loyo, J.M. et al. (2019). Design of an electric vehicle fast-charging station with integration of renewable energy and energy storage systems. *Electrical Power and Energy Systems* 105: 46–58.

6 Bai, S. and Lukic, S.M. (2013). Unified active filter and energy storage system for an MW electric vehicle Charging Station. *IEEE Transactions on Power Electronics* 12: 5793–5803.

7 Saccomanno, F. (2003). *Electric Power Systems Analysis and Control*, 1e. Wiley-IEEE Press.

8 Gnann, T., Funke, S., Jakobson, N. et al. (2018). Fast charging infrastructure for electric vehicles: today's situation and future needs. *Transportation Research Part D* 62: 314–329.

9 IEEE Std 2030.8-2018 (2018). *IEEE Standard for the Testing of Microgrid Controllers*. IEEE Standards.

10 Sivaraman, P., Sharmeela, C., and Elango, S. (2021). Load flow analysis for microgrids. In: *Microgrid Technologies* (ed. C. Sharmeela and P. Sivaraman), 177–196. Scrivener Publishing LLC.

11 IEC TR 62933-2-200 (2021). *Electrical Energy Storage (EES) Systems – Part 2-200: Unit Parameters and Testing Methods – Case Study of Electrical Energy Storage (EES) Systems in EV Charging Station with PV*. IEC.

12 El-Hawary, M.E. (2008). *Introduction to Electrical Power Systems*, 1e. Wiley-IEEE Press.

13 IEEE Std 2836-2021 (2021). *IEEE Recommended Practice for Performance Testing of Electrical Energy Storage (EES) System in Electric Charging Stations in Combination with Photovoltaic (PV)*. IEEE Standards.

14 Palanisamy, S., Shanmugasundaram, L., and Chenniappan, S. (2022). Energy Storage Systems for Smart Power Systems. Artificial Intelligence-based Smart Power Systems, 99–114. *Portico*. https://doi.org/10.1002/9781119893998.ch5.

15 Sharmeela, C., Sivaraman, P., Sanjeevikumar, P., and Holm-Nielsen, J.B. (Eds.) (2021). Microgrid Technologies. https://doi.org/10.1002/9781119710905.

8

Surge Protection Device for Electric Vehicle Fast-Charging Infrastructure

8.1 Introduction

The overvoltage and overcurrent in a low-voltage system is normally originated mainly from two sources called lightning and switching. The lightning overvoltage and overcurrent could either be due to direct lightning stroke to power system equipment and its installation like structures or indirect lightning stroke nearer to the power system equipment. Lightning is a natural phenomenon and it can significantly affect the power circuits and communication circuits. This lightning stroke affects the power system equipment and its operation in three ways [1, 2]. They are direct lightning stroke, indirect lightning near stroke, and indirect lightning far stroke. Figure 8.1 shows the direct and indirect lightning strokes on the power system equipment. Such surges if left unprotected can permanently damage the equipment and could potentially lead to irreparable damage.

Apart from surge protection devices (SPD), the fast-charging stations can be designed with a mast (i.e. lightning arrester) to protect the electric vehicle supply equipment (EVSE) from direct lightning strikes. Also, if the incoming power supply line is adequately protected through shield wires, impact from lightning strikes traveling through the power system can also be reduced. However, this chapter covers the requirement of SPD for fast-charging station rather than the design and need for mast/shield wire-based solution.

The switching overvoltage and overcurrent are generally due to intentional power system operations like capacitor bank switching, load switching, etc., and unintentional power system events like faults, short-circuits, etc. Normally, any of the above-mentioned power system events like switching operations, faults, etc., can result in switching overvoltages. The magnitude of overvoltage due to switching depends on

Fast-Charging Infrastructure for Electric and Hybrid Electric Vehicles: Methods for Large-Scale Penetration into Electric Distribution Networks, First Edition. Sivaraman Palanisamy, Sharmeela Chenniappan, and Sanjeevikumar Padmanaban.

Published 2023 by John Wiley & Sons, Inc.

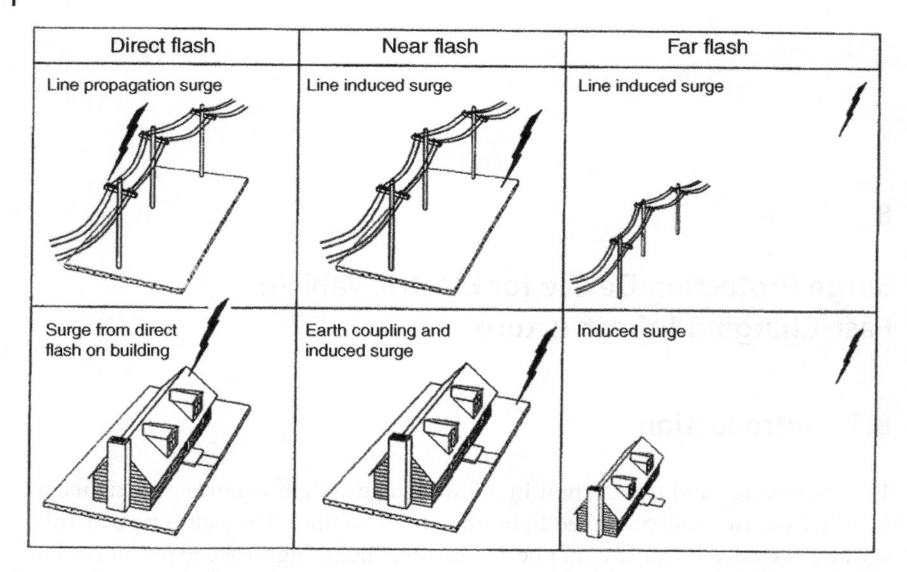

Figure 8.1 Direct and indirect lightning stroke on to the power system equipment [1] / with permission of IEEE.

circuit type, type of switching operation such as opening, closing, and restriking, loads, and the type of switching device.

Generally, fast-charging equipment are needed for public and/or private charging stations to charge the vehicles in a shorter duration [3–5]. The physical location of these public and private charging stations is typically classified into three types. In some locations, they can overlap each other to a certain extent based on actual site installation [6]. The type of site installation can also be potentially chosen based on the geographical location and the historical lightning activity in the area.

- Open locations: Locations are not covered or shelters are not provided. Some public charging stations or private charging stations come under this category. The open parking lots and the top level of the parking garages are the best example of this type. EVSE are subject to being affected by direct or indirect lightning strokes and impacts of overvoltage or overcurrent on EVSE are high.
- Covered locations: Locations are covered within the structure or under the roof. Some public charging stations and/or private charging stations come under this category. The enclosed parking garage is the best example of this type. EVSE are subject to get affected by indirect lightning strokes and impacts of overvoltage or overcurrent on EVSE are less as compared with open locations because during the direct lightning stroke to ground, conducting materials of a roof and its supporting structure will help to reduce and/or divert the overvoltage or overcurrent.

- Underground locations: Locations are within a structure below the ground level. Some public charging stations or private charging stations come under this category. The underground parking in an office premises is the best example of this type. EVSE are subject to get affected by indirect lightning strokes and impacts of overvoltage or overcurrent on EVSE are less because the locations are protected from a direct lightning stroke to the ground.

Based on the actual site location type, the EVSE is installed in an open location or inside the shelter. Since the EVSE is receiving an electric power supply from an external power grid, the EVSE and EVs are vulnerable to damage from overvoltage or overcurrent by lightning. Switching surges can impact the EVSE irrespective of the installation location as it always travels through the power system. The installation of SPDs at charging stations improves the equipment's survivability. Figure 8.2 shows the lightning current distribution from an indirect lightning stroke.

The IEEE Std C62.41.1 and IEEE Std C62.41.2 designated certain areas of low-voltage circuits by surge intensity or magnitude as Category A, Category B, and Category C [1, 2]. Category A has the lowest surge intensity while Category C has the highest surge intensity. Figure 8.3 shows the example for the category classifications.

Location Category A: Within the part of the installation at some distance from the service entrance.
Location Category C: External part of the building from some distance.
Location Category B: Between Category A and Category C.

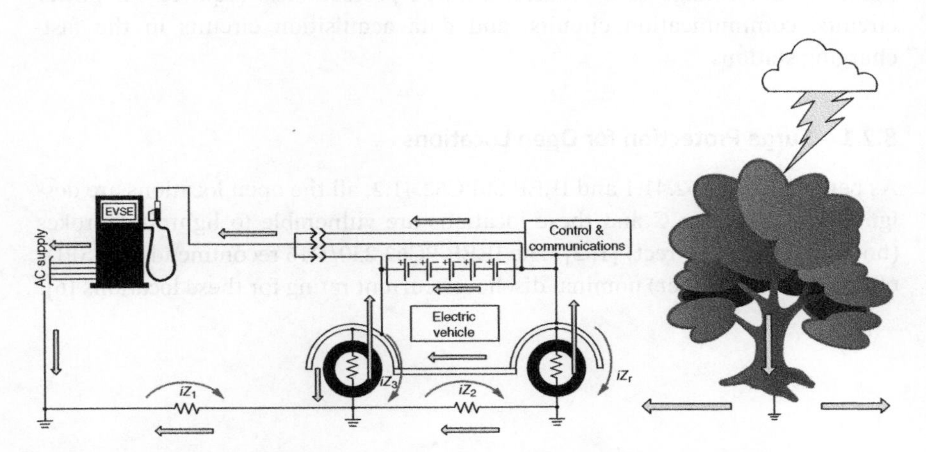

Figure 8.2 Simplified EVSE lightning current distribution from a nearby lightning ground flash [6] / with permission of IEEE.

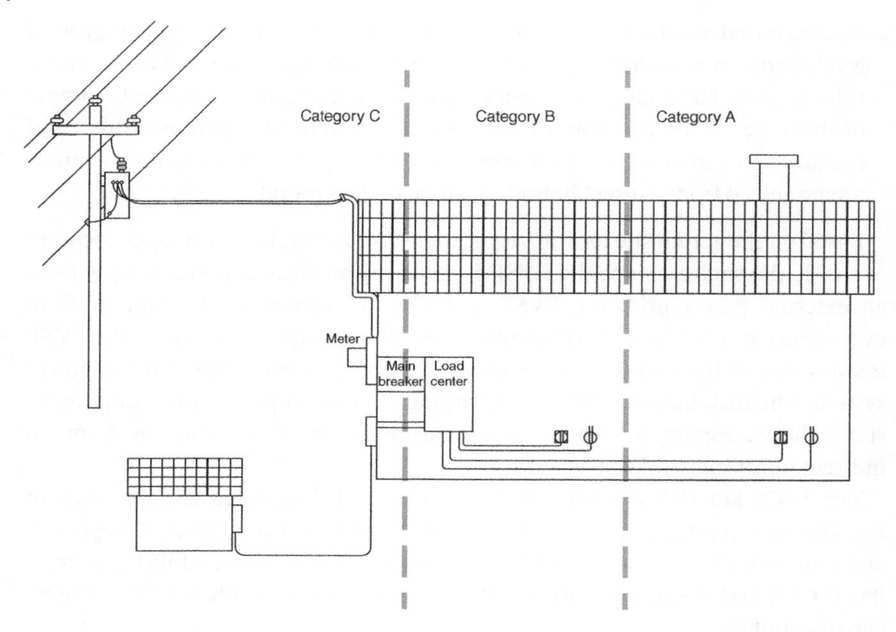

Figure 8.3 Example for the category classifications [6] / with permission of IEEE.

8.2 Surge Protection for Fast-Charging Stations

The SPDs are essential for a fast-charging station to prevent equipment from lightning overvoltage or overcurrent. Surge protection is required for power circuits, communication circuits, and data acquisition circuits in the fast-charging stations.

8.2.1 Surge Protection for Open Locations

As per IEEE Std C62.41.1 and IEEE Std C62.41.2, all the open locations are designated as Category C and these locations are vulnerable to lightning strokes (both direct and indirect) [1, 2]. The IEEE PC62.230/D16 recommends the SPDs of 10–20 kA (or higher) nominal discharge current rating for these locations [6].

Example 8.1 Surge protection for open location

The EVSE in these locations is mostly installed at outdoor premises. The SPD rating of 10–20 kA (or higher) nominal discharge current rating shall be installed at the incoming power source and EVSE as shown in Figure 8.4.

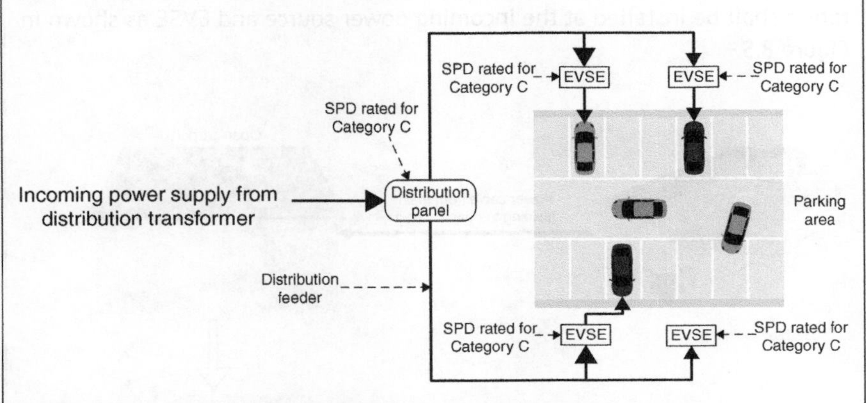

Figure 8.4 Example of category C SPD requirement for open location.

8.2.2 Surge Protection for Covered Locations

As per IEEE Std C62.41.1 and IEEE Std C62.41.2, all the covered locations are designated as Category C or Category B based on the construction and/or location of installation [1, 2].

Example for Category C: An open carport and small structure deriving the electricity from the nearby larger building.
Example for Category B: Garage attached to the building.

These locations are vulnerable to lightning strokes (both direct and indirect). The IEEE PC62.230/D16 recommends the SPDs of 10–20 kA (or higher) nominal discharge current rating for Category C locations and 3–10 kA nominal discharge current rating for Category B locations [6].

Example 8.2 Surge protection for covered location away from main building

The EVSE is installed in the open carport area. The carport is almost 100 m away from the main building and it receives the power supply from the main building. The SPD rating of 10–20 kA (or higher) nominal discharge current rating shall be installed at the incoming power source and EVSE as shown in Figure 8.5.

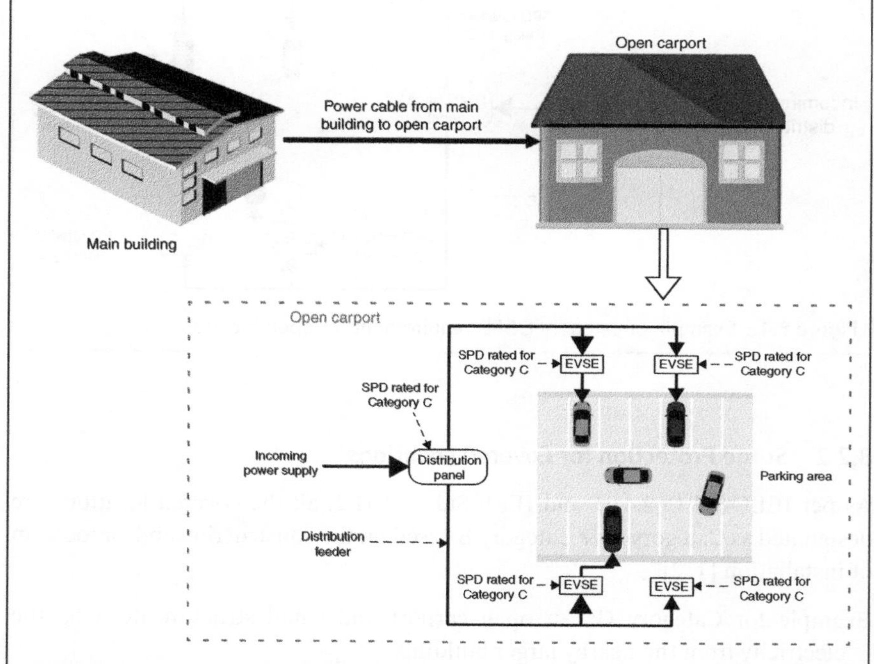

Figure 8.5 Example of category C SPD requirement for covered location away from main building.

Example 8.3 Surge protection for covered location near to main building

The EVSE in the garage is attached to the building. The SPD rating of 10–20 kA (or higher) nominal discharge current rating shall be installed at the incoming power source and the 3–10 kA nominal discharge current rating installed at the EVSE as shown in Figure 8.6.

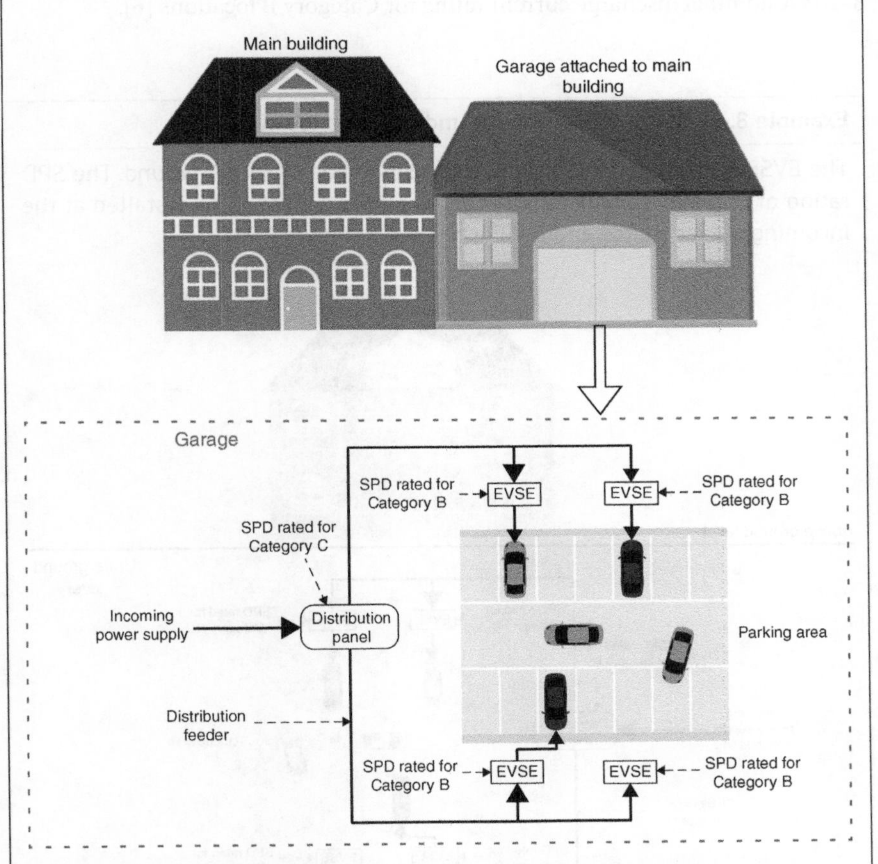

Figure 8.6 Example of category C SPD requirement for covered location near to main building.

8.3 Surge Protection for Underground Locations

The underground location is shielded from direct lightning stroke and hence these locations are considered as Category B locations. These locations are vulnerable to indirect lightning strokes and they can affect the EVs and vehicles in the underground locations. The IEEE PC62.230/D16 recommends the SPDs of 3–10 kA nominal discharge current rating for Category B locations [6].

Example 8.4 Surge protection for underground location

The EVSE in these locations is mostly installed at the underground. The SPD rating of 3–10 kA (nominal discharge current rating) shall be installed at the incoming power source and EVSE as shown in Figure 8.7.

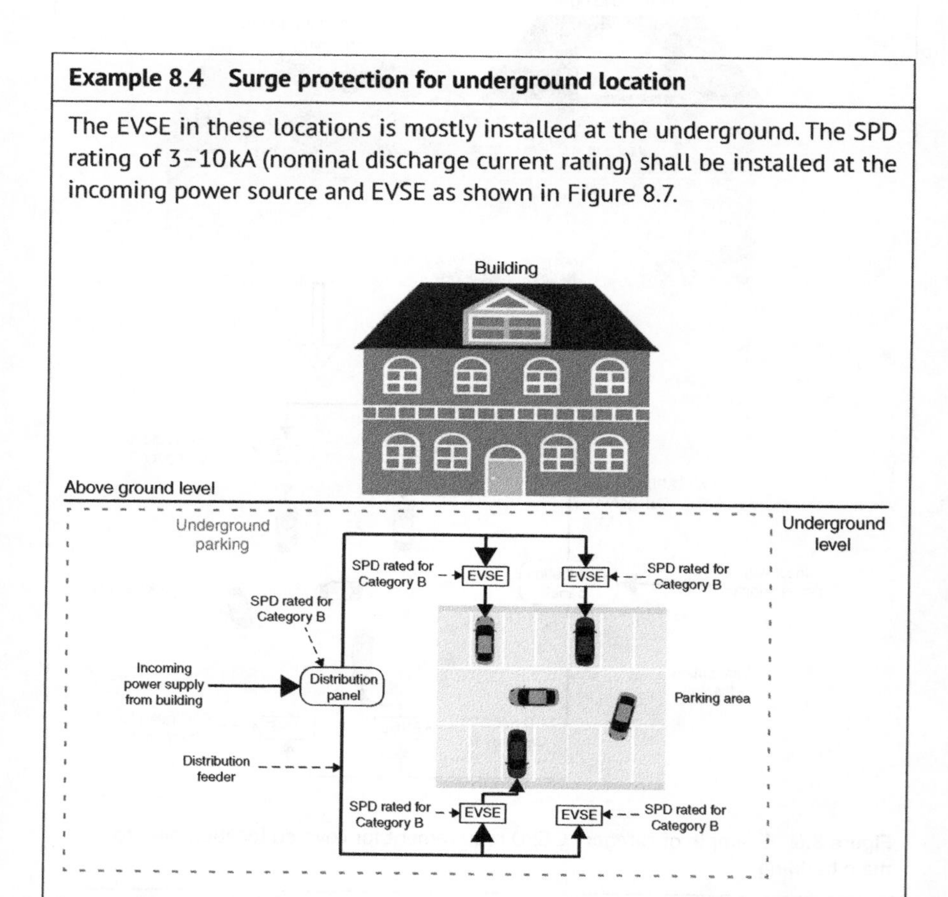

Figure 8.7 Example of category B SPD requirement for underground location.

8.4 Conclusion

Lightning is a natural phenomenon and it will create overvoltage or overcurrent in an electric power system. SPD are widely used to prevent electrical equipment from overvoltage and overcurrent produced by lightning strokes. The electric vehicle fast-charging station equipment are also subject to the risk of overvoltage and overcurrent produced by direct or indirect lightning stroke. The SPD is essential and needs to be installed in the fast-charging station to prevent the EVSE from overvoltage and overcurrent. This chapter discussed the requirements of SPD for the fast-charging station in open locations, covered locations, and underground locations.

References

1 IEEE Std C62.41.1-2002 (2002). *IEEE Guide on the surge environment in low-voltage (1000 V and Less) AC power circuits*. IEEE Standards.

2 IEEE Std C62.41.2-2002 (2002). *IEEE recommended practice on characterization of surges in low-voltage (1000 V and less) AC power circuits*. IEEE Standards.

3 Sivaraman, P., Sharmeela, C., and Logeshkumar, S. (2021). Charging infrastructure layout and planning for plug-in electric vehicles. In: *Cable Based and Wireless Charging Systems for Electric Vehicles* (ed. P. Sanjeevikumar), 1–24. IET.

4 Sivaraman, P. and Sharmeela, C. (2021). Power quality problems associated with electric vehicle charging infrastructure. In: *Power Quality in Modern Power Systems* (ed. P. Sanjeevikumar, C. Sharmeela, and J.B. Holm-Nielsen). London: Academic Press.

5 Ministry of Power (2022). *Charging Infrastructure for Electric Vehicles – Guidelines and Standards*, 14 January, 1–7. New Delhi, India: Ministry of Power https://powermin.gov.in/sites/default/files/Final_Consolidated_EVCI_Guidelines_January_2022_with_ANNEXURES.pdf (accessed 12 March 2023).

6 IEEE PC62.230/D16 Draft Guide for Surge Protection of Electric Vehicle Infrastructure, May 2022. IEEE standard

9

Power Quality Problems Associated with Fast-Charging Stations

9.1 Introduction

The fast chargers in a charging station are used to charge the EVs in less time duration [1]. These fast chargers draw high power from the grid and are used to charge the EVs. The main advantage of using fast chargers is reducing the waiting time of EV users while charging at public charging stations. At the same time, fast chargers have a negative impact on the power system by introducing many power quality (PQ) problems. Fast chargers are made up of power electronics-based elements and this will introduce the PQ issues such as harmonics, DC injection, and flicker into the power system [2]. The EV charging loads are not uniform throughout the charging process, i.e. power flow from the charger to the EV is not uniform and it varies based on the battery state of charge (SoC) [3]. Also, it is not possible to predict the exact amount of load connected at any point in time to the charging stations. In simple words, load impedance (i.e. battery impedance) varies over any given charging period and hence the current flow also varies. These EV charging loads are connected across the distribution system, and the system impedance also changes throughout the charging process, resulting in changes in the system resonance point.

Modern power electronics components in the fast chargers are switching in high frequency in the kilohertz range. The exact switching frequency and logic is also not consistent and depends upon the technology implemented by the manufacturers. In general, high-frequency switching introduces another PQ problem called supra-harmonics.

This chapter discusses the PQ problems such as current harmonics, DC injection, flicker, system resonance, and supra-harmonics.

Fast-Charging Infrastructure for Electric and Hybrid Electric Vehicles: Methods for Large-Scale Penetration into Electric Distribution Networks, First Edition. Sivaraman Palanisamy, Sharmeela Chenniappan, and Sanjeevikumar Padmanaban.

Published 2023 by John Wiley & Sons, Inc.

9.2 Introduction to Power Quality

The IEC standard 61000-4-30:2015 defines PQ as "characteristics of the electricity at a given point on an electrical system, evaluated against a set of reference technical parameters" [4]. The IEEE standard 1100-2005 defines PQ as "the concept of powering and grounding electronic equipment in a manner that is suitable to the operation of that equipment and compatible with the premise wiring system and other connected equipment" [5].

The IEEE standard 1159-2019, IEEE recommended practice for monitoring electric PQ categorize the PQ phenomenon into transients, short duration RMS variation, long duration RMS variation, imbalance, waveform distortion, voltage fluctuations, and power frequency fluctuations [6]. The typical magnitude and duration of this PQ phenomenon as per IEEE 1159-2019 is listed in Table 9.1.

Transients: The transients are momentary changes in voltage or current or both due to faults, switching operations, lightning, capacitor bank switching, etc.

Impulsive transients: Impulsive transients are sudden change in voltage or current or both in unidirectional, either positive or negative.

Oscillatory transients: Oscillatory transients are sudden change in voltage or current or both in bidirectional, i.e. changes in both positive and negative polarity.

Voltage sag: The voltage sag is short duration reduction of RMS voltage magnitude for shorter time period. IEEE 1159-2019 defines the voltage sag as decrease in RMS voltage magnitude between 0.9 and 0.1 pu for the time duration of 0.5 cycles to less than one minute.

Voltage swell: The voltage swell is short duration increase of RMS voltage magnitude for shorter time period. IEEE 1159-2019 defines the voltage swell as increase in RMS voltage magnitude between 1.1 and 1.8 pu for the time duration of 0.5 cycles to less than one minute.

Momentary interruption: Momentary interruptions are reductions in RMS voltage less than 0.1 pu for the time duration of less than one minute.

Sustained interruption: Sustained interruptions are reductions in RMS voltage less than 0.1 pu for a time duration of more than one minute.

Voltage unbalance: Voltage unbalance in a three-phase system is the RMS values of the voltages and/or the phase angles among the three phases are not equal.

Voltage harmonics: Voltage harmonics are the presence of frequencies that are integral multiples of the fundamental frequency (50 or 60 Hz) of voltage.

Current harmonics: The current harmonics are the presence of frequencies that are integral multiples of the fundamental frequency (50 or 60 Hz) of current.

Inter-harmonics: The inter-harmonics are the presence of frequencies that are not an integral multiple of the fundamental frequency (50 or 60 Hz) of voltage or current.

Table 9.1 Typical magnitude and duration of the power quality phenomenon [6] / with permission of IEEE.

Categories	Typical spectral content	Typical duration	Typical voltage magnitude
1.0 Transients			
1.1 Impulsive			
1.1.1 Nanosecond	5 ns rise	< 50 ns	
1.1.2 Microsecond	1 μs rise	50 ns–1 ms	
1.1.3 Millisecond	0.1 ms rise	>1 ms	
1.2 Oscillatory			
1.2.1 Low frequency	<5 kHz	0.3–50 ms	0–4 pu[a]
1.2.2 Medium frequency	5–500 kHz	20 μs	0–8 pu
1.2.3 High frequency	0.5–5 MHz	5 μs	0–4 pu
2.0 Short-duration root-mean-square (rms) variations			
2.1 Instantaneous			
2.1.1 Sag		0.5–30 cycles	0.1–0.9 pu
2.1.2 Swell		0.5–30 cycles	1.1–1.8 pu
2.2 Momentary			
2.2.1 Interruption		0.5 cycles–3 s	<0.1 pu
2.2.2 Sag		30 cycles–3 s	0.1–0.9 pu
2.2.3 Swell		30 cycles–3 s	1.1–1.4 pu
2.2.4 Voltage imbalance		30 cycles–3 s	2–15%
2.3 Temporary			
2.3.1 Interruption		>3 s–1 min	<0.1 pu
2.3.2 Sag		>3 s–1 min	0.1–0.9 pu
2.3.3 Swell		>3 s–1 min	1.1–1.2 pu
2.3.4 Voltage imbalance		>3 s–1 min	2–15%
3.0 Long-duration rms variations			
3.1 Interruption, sustained		>1 min	0.0 pu
3.2 Undervoltages		>1 min	0.8–0.9 pu
3.3 Overvoltages		>1 min	1.1–1.2 pu
3.4 Current overload		>1 min	
4.0 Imbalance			
4.1 Voltage		Steady state	0.5–5%
4.2 Current		Steady state	1.0–3.0%

(Continued)

Table 9.1 (Continued)

Categories	Typical spectral content	Typical duration	Typical voltage magnitude
5.0 Waveform distortion			
5.1 DC offset		Steady state	0–0.1%
5.2 Harmonics	0–9 kHz	Steady state	0–20%
5.3 Interharmonics	0–9 kHz	Steady state	0–2%
5.4 Notching		Steady state	
5.5 Noise	Broadband	Steady state	0–1%
6.0 Voltage fluctuations	<25 Hz	Intermittent	0.1–7%
			0.2–2 P_{st} [b]
7.0 Power frequency variations		<10 s	±0.10 Hz

These terms and categories apply to power quality measurements and are not to be confused with similar terms defined in IEEE Std 1366™-2012 [B30] and other reliability-related standards, recommended practices, and guides.

[a] The quantity *pu* refers to *per unit*, which is dimensionless. The quantity 1.0 pu corresponds to 100%. The nominal condition is often considered to be 1.0 pu. In this table, the nominal peak value is used as the base for transients and the nominal rms value is used as the base for rms variations.

[b] Flicker severity index Pst as defined in IEC 61000-4-15:2010 [B17] and IEEE Std 1453™ [B31].

Voltage fluctuations: The voltage fluctuations are a continuous change in the instantaneous cycle-to-cycle voltage due to the variations of load resistance in every cycle.

Rapid voltage change: A quick transition in RMS voltage occurring between two steady-state conditions, and during which the RMS voltage does not exceed the dip(sag)/swell thresholds. In simple words, steady-state voltage variation within the steady-state tolerance limits.

Supra-harmonics: Supra-harmonics are generally termed as the harmonic spectrum content in the range of 2–150 kHz.

9.3 Power Quality Problems Due to Fast-Charging Stations

The PQ comprises two aspects, i.e. voltage quality and current quality [7]. If anyone one of the parameters is affected or deviated from the reference value, the overall PQ becomes poor.

The voltage quality parameters are:

- Voltage sag and swell
- Interruptions
- Voltage unbalance
- Voltage fluctuations
- Transients
- Voltage harmonics

These voltage quality parameters mostly originate from the grid side (source side), from the load side where the extensive use of voltage source converters has an impact on the connected equipment in the charging station.

The current quality parameters are:

- Current harmonics
- DC injection
- Current unbalance

These current quality parameters mostly originate from the load side (i.e. end-user equipment) and this will have an impact on the grid.

Hence, maintaining the quality of power which is required for reliable and trouble-free operation of end users as well as utility equipment shall require the combined effort from end users and utility. The PQ problems of fast-charging stations can be seen in two ways [2].

1) Impact of poor PQ of distribution grid on fast-charging station loads.
2) Impact of poor PQ from the fast-charging station loads on the distribution grid.

9.3.1 Impact of Poor Power Quality of Distribution Grid on Fast-Charging Station Loads

The fast-charging stations receive the electric power supply from the DISCOMs distribution system. If the DISCOMs incoming supply quality is poor, then the connected fast-charging station loads get affected.

The distribution grid supplies the power to the connected loads in the charging stations and it is the DISCOMs responsibility to provide high-quality incoming supply voltage to the loads. Due to various reasons, the voltage quality of the distribution system gets affected. For example, any faults in the transmission system and/or distribution system results in interruption or voltage sag in the distribution system, and it will have an impact on the performance of the charger [8, 9]. Also, the connection of more power electronic devices in the vicinity of the charging station will introduce more harmonics into the distribution system, which could also negatively impact the performance of the fast-charging loads. The following voltage quality parameters originate in the distribution system [20, 21], and they will have an impact on the equipment performance:

- Voltage sag and swell
- Interruptions
- Voltage unbalance
- Voltage fluctuations
- Transients
- Voltage harmonics

The SAE J2894-1 Power Quality Requirements for Plug-In Electric Vehicle Chargers provides the PQ requirements of the charger for the voltage sag [10]. Some of them are [11]:

- During the voltage sag, EV chargers have to remain energized for voltage reduced to 80% of nominal voltage for up to two seconds.
- EV chargers shall have fault ride-through capability to ride through the complete loss of voltage during a fault for up to 12 seconds.

9.3.2 Impact of Poor Power Quality from the Fast-Charging Station Loads on the Distribution Grid

The fast chargers in a charging station draw power from the grid and are used to charge the EVs in less time duration. This reduces the waiting time of EV users at public charging stations. At the same time, it has a negative impact on the power system by creating many PQ problems. The EV chargers are made up of power electronics-based components and this will introduce the PQ issues such as harmonics, DC injection, and flicker into the power system. The amount of current harmonic injection or emission from the EV charger is based on the following factors [11]:

- Power level or rating
- Charger circuit or topology
- Grid incoming voltage background voltage harmonic distortion
- Network impedance between the charger and measurement point

The EV charging loads are not uniform throughout the charging process, i.e. power flow from the charger to the EV is not uniform and it varies based on the battery SoC. In a single charging cycle, the EV charger will operate in two modes, namely constant current (CC) and constant voltage (CV). From zero to 80% or 85% of the battery SoC charger is in CC mode and after that, the charger will follow the CV mode until the battery is fully charged. In simple words, load impedance (i.e. battery impedance) varies during the charging period and hence the current flow also varies. These EV charging loads are connected across the distribution system, and the system impedance also changes throughout the charging process, resulting in changes in the system resonance point.

The power electronics components in the fast chargers utilize high-frequency switching in the kilohertz range [12]. High-frequency switching introduces another PQ problem called supra-harmonics [13].

9.4 Analysis of Harmonic Injection into the Distribution System

The primary and secondary distribution system provides the electric power supply to its customers within its boundary. The distribution system provides the power supply to its customers and their boundaries are typically in the range of 3–8 km^2. These distribution systems are powering the residential, commercial, and/or industrial power system loads. Due to the sophisticated operation and control requirements, many equipment in residential, commercial, and industrial systems are using power electronics-based circuits. Nowadays, solar PV systems are integrating into the distribution system as a distributed generation. These solar PV systems are using power electronics-based components for DC to AC conversion. The characteristics of these power electronics-based equipment are nonlinear in nature. During their operation, power electronics-based equipment injects current harmonics into the system [14].

The current harmonic injection from these equipment is not always uniform and the harmonic current magnitude depends on how many equipment are in operation at that time. In addition to the existing loads in the distribution system, EV charging loads get connected across the distribution system.

Whenever the significant number of nonlinear loads or equipment are typically in the range of 25–30% or higher than the total load and if any equipment failure are reported, it is essential to perform harmonic analysis to find out the harmonic magnitude and its spectrum for different operating scenarios. The IEEE Std 3002.8-2018 lists the following calculation methods to investigate and determine the different voltage/current harmonic spectrum [15]:

i) Hand calculation or manual calculation
ii) Conducting field measurements at the site
iii) Model calibration
iv) Computer simulation

Sometimes for assessment and compliance verification, a combination of the above methods needs to be adopted or can be effectively adopted. As a case, for example, conducting field measurements at the site can help in validating or fine-tuning the harmonic model of the charging system. The validated model after thorough investigation and agreement with the corresponding manufacturer can be utilized for simulations of similar upcoming charging docks to analyze the impact and take appropriate harmonic mitigation measures as required.

9.4.1 Hand Calculation or Manual Calculation

The voltage and current harmonics can be analyzed by means of manual calculations or hand calculations for a smaller distribution network (generally, radial distribution system). However, this method is not suitable for large distribution networks (which are potentially meshed and/or have significantly long lines) due to the tedious calculation involved and susceptible to manual errors. Though this method can give a good first-hand overview of the distortion that we could expect, it will not be generally considered satisfactory for getting approval for a charging station during the design phase. For such approvals, typically an industry-wide accepted simulation tool might be required to validate and highlight the acceptance of the harmonic spectrum.

9.4.2 Conducting Field Measurements at the Site

The voltage and current harmonics in the system can be analyzed using a PQ analyzer through field measurements at various locations in the distribution network. This method is more suitable for the distribution system with the existing loads and this method is not suitable to analyze the voltage and current harmonics due to future anticipated loads. The IEEE Std 519-2022 provides the recommended harmonic limits (both voltage and current harmonics) at the point of common coupling (PCC) and the PQ analyzer used for the measurement shall be Class A as per IEC 61000-4-30 [16].

The typical single line diagram (SLD) for the location of PQ measurement for a fast-charging station for compliance with harmonic limits as per IEEE 519-2022 is shown in Figure 9.1. Also, this method can be used to measure the voltage and current harmonics along with other PQ parameters at any location as per the requirements.

From the measurements, the various PQ issues including harmonics at the measurement points on the prevailing system loading/operating conditions can be analyzed. The accuracy of the measured data is subjected to the accuracy of the measuring instruments such as the PQ analyzer, current transformer (CT), and potential transformer (PT). This method effectively provides the voltage and current harmonics along with various PQ issues for the entire charging duration (i.e. complete charging cycle).

The major advantage of site measurement is that the exact harmonic levels at the point of connection are understood and careful future planning can be

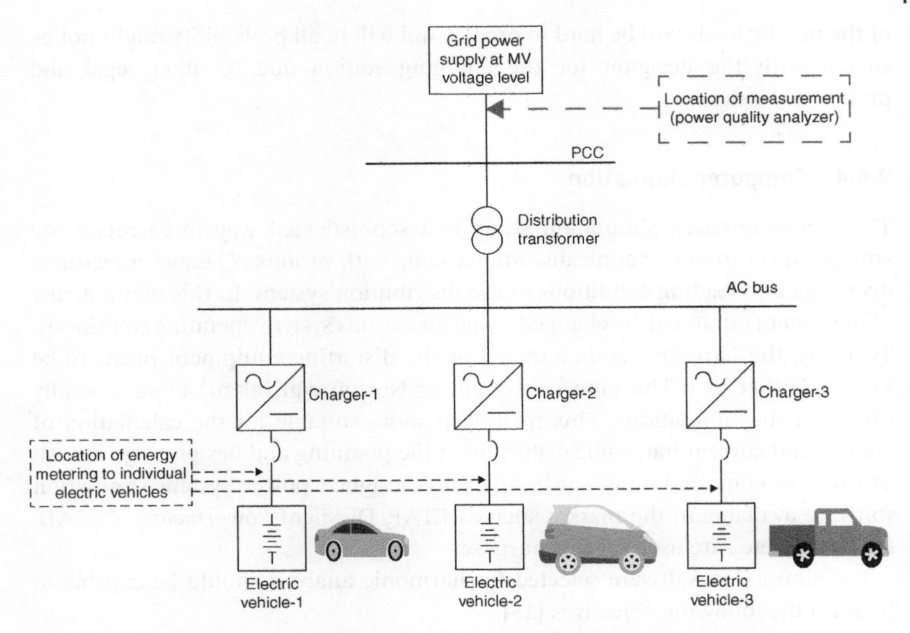

Figure 9.1 SLD for the location of power quality measurement for a fast-charging station.

undertaken. Site measurements have the potential to even cover potential scenarios which might have been overlooked by simulations during the design phase. Site measurements also give a much deeper insight into the design offered for the charging station equipment. The disadvantage of this method is that it is not suitable for the future addition of charging loads and is not possible to analyze the voltage and current harmonics for different grid operating conditions as continuous site measurements over an extended period are not feasible to process and analyze.

9.4.3 Model Calibration

Model calibration is one of the approaches to determine the voltage and current harmonics distortion in the distribution system for the various operating and loading conditions created in the test bed. This method shall be used to predict the voltage and current harmonics of the distribution system, especially during the planning and design stage. If any of the harmonic orders is exceeding the limit, the suitable mitigation measures and the rating required to maintain the harmonic distortions within the limits can be analyzed. The results of this method shall be in close agreement with actual system performance. The hardware-in-loop is widely used to create the test bed. The major disadvantage is that the exact nature of the nearby loads will be hard to predict and will in all probability might not be shared with the designer for the charging station due to other legal and proprietary issues.

9.4.4 Computer Simulation

The computer-based simulation study is a sophisticated way to calculate the voltage and current harmonic distortions along with various PQ issues for various operating and loading conditions of the distribution system. In this method, any of the system inputs can be changed easily for various system operating conditions. Typically, the harmonic source model of the distorting equipment needs to be known (either as a Thevenin equivalent or Norton equivalent) to successfully carry out the simulations. This method is more suitable for the calculation of voltage and current harmonic distortion at the planning and design stage as well as the post-commissioning stage. Specially designed power system simulation software available in the market such as ETAP, Digsilent Powerfactory, PSCAD, PSS Sincal, etc., are used for this purpose.

The simulation software selected for harmonic analysis should be capable to perform the following objectives [15]:

- Voltage harmonic distortion at various buses
- Current harmonic distortion at various branches
- Individual and total voltage harmonic distortion
- Individual and total current harmonic distortion
- Impedance vs frequency spectrum (for identifying the system resonance conditions)
- Alert for equipment overloading or overstress

The step-by-step procedure for performing harmonic analysis for the distribution system by means of simulation software is given below.

Step 1: Up-to-date SLD of the distribution system indicating the cables, lines, transformers, capacitor banks, nonlinear loads, etc.

Step 2: Collect the equipment data including the harmonic model for the corresponding equipment and its ratings for all equipment.

Step 3: Collect the location of nonlinear loads including EV charger and harmonic spectrum of each nonlinear loads.

Step 4: Collect the maximum and minimum fault level at grid, system impedance, and background harmonic details for different system operating conditions. The grid impedance can be made available in various forms like either a polygon, sector, or multiple impedance snapshots each depicting an operating condition or maybe modeled just as a simple impedance range based on the short-circuit level. Modeling based on the short-circuit level generally does not offer very good insights into the possible system resonance condition.

Step 5: Collect the permissible harmonic distortion limits at PCC.

Step 6: Model the distribution system and its equipment's detail in power system simulation software such as ETAP, Digsilent Powerfactory, etc.

Step 7: Perform the harmonic analysis for the distribution system base configuration.

Step 8: Compute the total and individual voltage and current harmonic distortion at the PCC.

Step 9: Evaluate the results, i.e. compare the total and individual voltage and current harmonic distortion with permissible limits at PCC.

Step 10: If any harmonic order or total harmonic distortion value exceeded the limit, suitable harmonic mitigation devices are to be considered or modify the system details and perform the computation till all the values are within the limit.

These 10 steps provide the voltage and current harmonic values and their compliance for the distribution system base or normal operating configuration. The same steps are to be followed for other system operating conditions or topologies.

9.5 Analysis of System Resonance Condition

The system resonance is a condition in a power system and this condition can occur whenever the system inductive reactance is equal to capacitive reactance at some harmonic frequency. At this time, cyclic energy transfers between the capacitive elements and inductive elements. The frequency at which the resonance condition occurs is called resonant frequency [17].

The expression for system resonant condition is,

$$X_L = X_C \tag{9.1}$$

where,

X_L is inductive reactance
X_C is capacitive reactance

The resonant frequency can be obtained from,

$$F_r = \frac{1}{2\pi\sqrt{LC}} \tag{9.2}$$

where,

F_r is resonant frequency
L is system inductance
C is system capacitance

The resonance conditions are classified into two types:

1) Series resonance
2) Parallel resonance

The series resonance condition occurs whenever the system's inductive reactance and capacitive reactance are in series. In this condition, the impedance offered by the system is low as shown in Figure 9.2 and resulting in unexpected current harmonic flow through the equipment. Single tuned shunt filters designed as one of the mitigation measures for harmonics follows the principle of series resonance and thereby acting as sink for the harmonic generated in the system.

The parallel resonance condition occurs whenever the inductor is parallel with a capacitor. In this condition, the impedance offered by the system is high as shown in Figure 9.3 and results in excessive voltage harmonic distortion across the equipment.

Whenever the harmonics in the distribution system cause any problems like cable failure or overloading, capacitor bank failure, etc., it is essential to find the resonant frequencies at the site. A parallel system resonance condition may occur at or closer to lower harmonic orders like 3rd, 5th, 7th, etc. This resonance point becomes critical if many nonlinear loads have significant current harmonics injection in that frequency range.

The EV charging loads are a new type of load in the power system. During the charging period, the load impedance (i.e. battery impedance) will vary with respect to the SoC of the battery. In simple words, from a grid perspective, the connected load impedance of the EV loads is not uniform, and it will vary throughout the charging process. This change in load impedance may change the system resonance

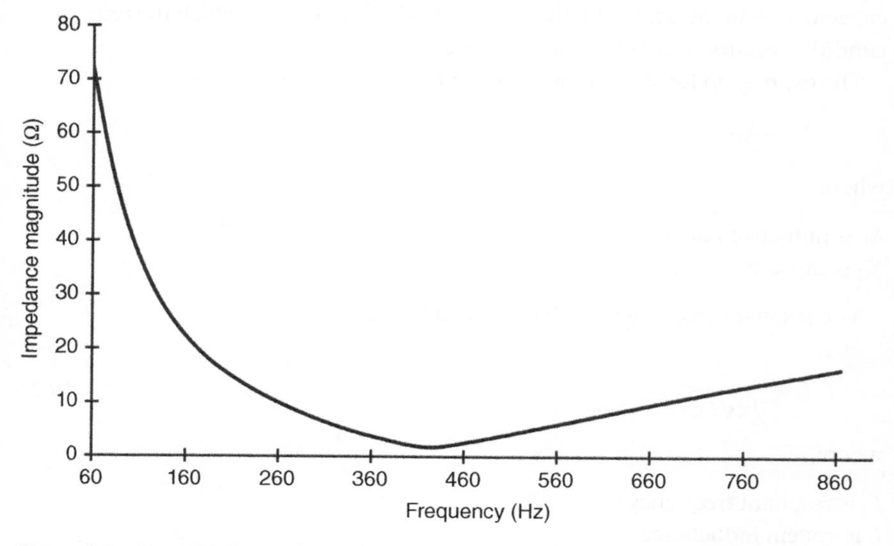

Figure 9.2 Impedance vs frequency characteristics for series resonance [15]/ with permission of IEEE.

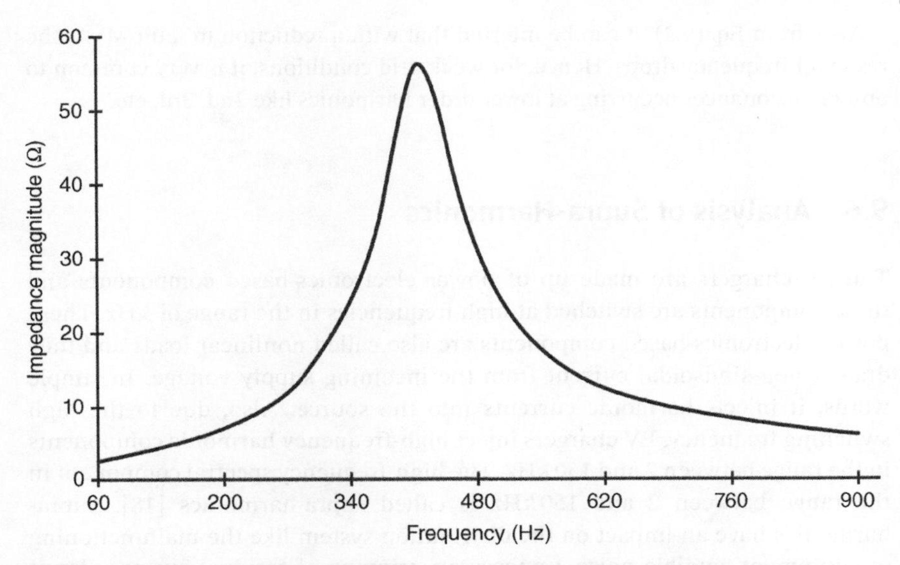

Figure 9.3 Impedance vs frequency characteristics for parallel resonance [15]/ with permission of IEEE.

point to multiple points during the single EV charging process. It becomes a difficult task when there is large penetration of EV charging loads that are connected across the distribution system. Multiple EVs may charge at the same time through the same distribution system and the connected load impedance highly varies, and it will result in constant change in the system resonance point. If any nonlinear load including existing nonlinear loads in the distribution system and EV charging loads injects harmonics at or closer to the resonant frequency, amplification of the voltage/current can be significant, and this leads to excessive current harmonic flow in branch circuits elements or voltage harmonic distortion at various busses.

It is essential to mitigate the impact of system resonance conditions. This can be achieved by adding the reactor in series with the capacitor. Adding a reactor will shift the system to inductive reactance at the resonant frequency and capacitive reactance at the fundamental frequency. The resonant frequency of the system in the addition of the capacitor bank can be obtained from Eq. (9.3).

$$F_r = F_{\text{fund}} X \sqrt{\frac{\text{Fault MVA}}{\text{MVAR}}} \tag{9.3}$$

where,

F_r is resonant frequency

F_{fund} is the fundamental frequency (i.e. 50 or 60 Hz)

Fault MVA is the short-circuit capacity of the system in MVA

MVAR is a capacitor bank rating

Also, from Eq. (9.3), it can be inferred that with a reduction in fault MVA, the resonant frequency drops. Hence, for weak grid conditions, it is very common to observe resonances occurring at lower order harmonics like 2nd, 3rd, etc.

9.6 Analysis of Supra-Harmonics

The EV chargers are made up of power electronics-based components and these components are switched at high frequencies in the range of kHz. These power electronics-based components are also called nonlinear loads and they draw a non-sinusoidal current from the incoming supply voltage. In simple words, it injects harmonic currents into the source. Also, due to the high switching frequency, EV chargers inject high-frequency harmonic components in the range between 2 and 150 kHz. The high-frequency spectral component in the range between 2 and 150 kHz is called supra-harmonics [18]. Supra-harmonics have an impact on the distribution system like the malfunctioning of equipment, audible noise, unnecessary tripping of residual current circuit breakers, overheating, interference with communication devices, errors in meter readings, etc.

The draft version of IEC 61000-4-30:2022 gives the requirements of metering instruments used for supra-harmonics. Annex C of the draft IEC 61000-4-30:2022 gives the functional design and specification for measurements in the range of 2–9 kHz for Class A and Class S equipment. Also, Annex D of the same draft gives the functional design and specifications for measurements in the range of 9–150 kHz given as informative [19]. The PQ analyzer or harmonic analyzer used for measurement of supra-harmonics shall have the features specified in Annex C and Annex D of IEC 61000-4-30:2022.

9.7 Case Study: Harmonic Measurement of 30 kW DC Fast Charger

The harmonic measurement has been carried out for the public fast-charging station shown in Figure 9.4. The harmonics produced by the EV chargers are measured for 30 kW DC fast charger. The Dranetz HDPQ Visa power quality analyzer, class A as per IEC 61000-4-30 is used for the harmonic assessment at the site location as shown in Figure 9.4, i.e. Dranetz HDPQ Visa – power quality analyzer is connected at the terminal of AC side of the charger for harmonic measurement.

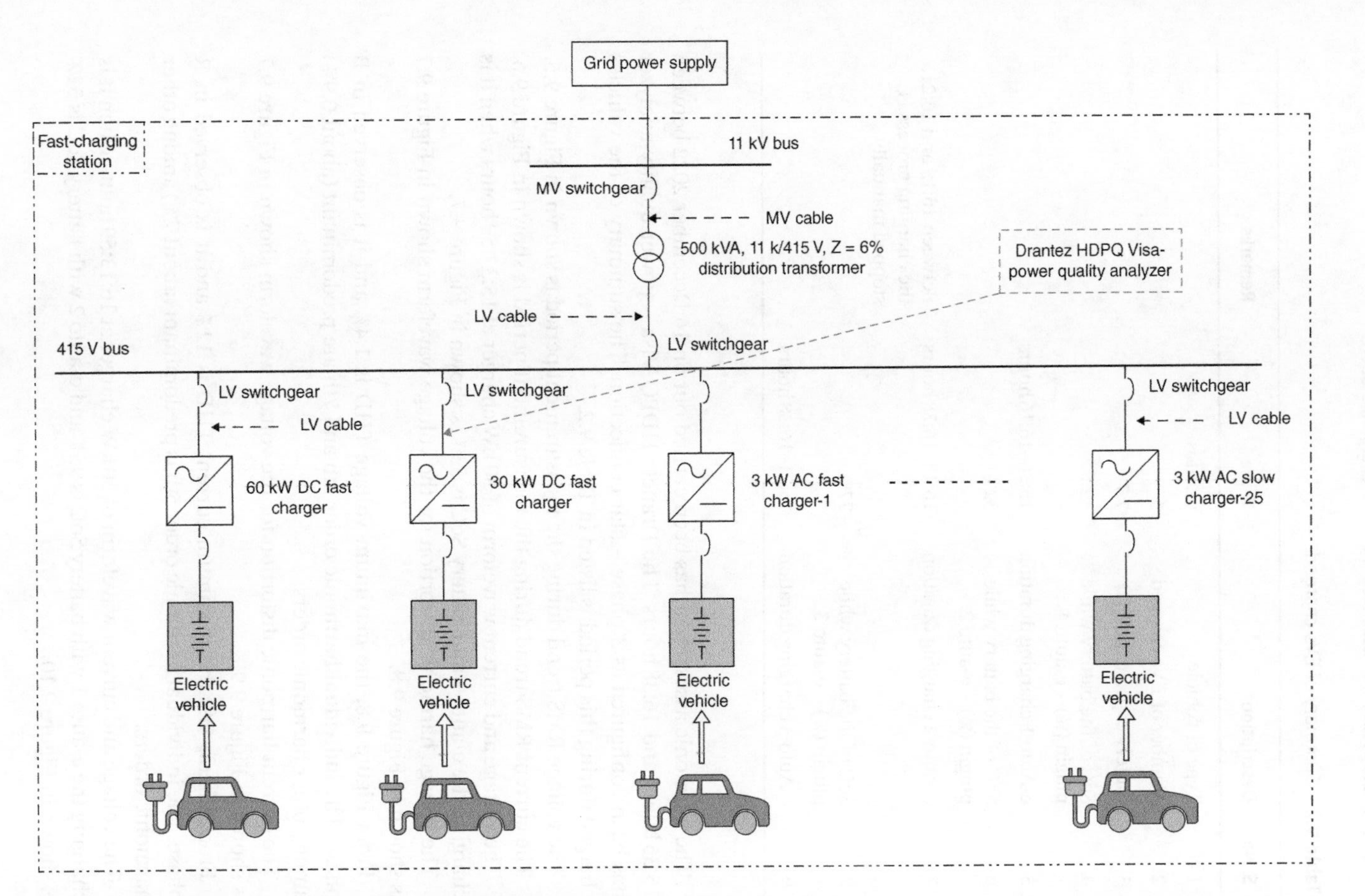

Figure 9.4 Location of harmonic measurement of EV chargers.

Table 9.2 Summary of EVs charged.

S. No.	Description	Value	Remarks
1	Type of vehicle	e-autos	
2	Number of EVs charged	3	
3	Battery capacity (kWH)	7.7	
4	SoC of the battery while plugin (%) – e-auto 1	52%	
5	e-Auto 1 charging duration	15:36–16:16 hours	
6	SoC of the battery while plugin (%) – e-auto 2	32%	
7	e-Auto 2 charging duration	16:13–16:38 hours	Between 16:15 and 16:21, the charging process is stopped manually
8	SoC of the battery while plugin (%) – e-auto 3	77%	
9	e-Auto 3 charging duration	16:21–16:38 hours	

The harmonic assessment has been carried out on 16 December 2022 between 15:36 hours and 16:38 hours. The Dranetz HDPQ Visa – power quality analyzer has been configured as 3 phase – star connection. The summary of the vehicles charged during this period is listed in Table 9.2.

The voltage RMS trend during the measurement period is shown in Figure 9.5.

The current RMS trend during the measurement period is shown in Figure 9.6.

The voltage and current waveform of 30 kW charger at 15:37:59 hours when it is charging the e-auto 1 and battery SoC is 53% is shown in Figure 9.7.

The voltage harmonic distortion for the voltage waveform shown in Figure 9.7 is shown in Figure 9.8.

From Figure 9.8, the maximum voltage THD is 1.4% and it is observed in B phase. The individual harmonic order 5th and 7th are predominant (about 0.9%) among other harmonic orders.

The current harmonic distortion for the voltage waveform shown in Figure 9.7 is shown in Figure 9.9.

From Figure 9.9, the maximum current THD is 12% and it is observed in R phase. The individual harmonic order 5th is predominant (about 7%) among other harmonic orders.

The voltage and current waveform of 30 kW charger at 16:13:59 hours when it is charging the e-auto 1 with battery SoC is 95% and e-auto 2 with battery SoC is 32% is shown in Figure 9.10.

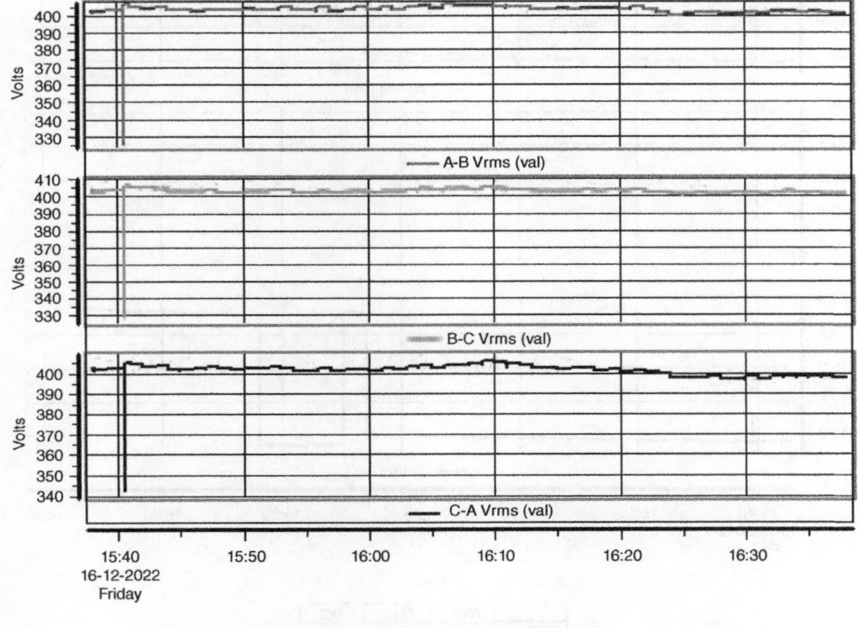

Event #1 at 16-12-2022 15:36:48:758
Pre-trigger

	Min	Max	Avg
A-BVrms	325.9	406.6	401.5
B-CVrms	329.3	406.4	400.9
C-AVrms	342.3	406.2	400.7

Figure 9.5 Voltage RMS trend during the measurement period.

The voltage harmonic distortion for the voltage waveform shown in Figure 9.10 is shown in Figure 9.11.

From Figure 9.11, the maximum voltage THD is 1.6% and it is observed in B phase. The individual harmonic order 7th is predominant (about 1.1%) among other harmonic orders.

The current harmonic distortion for the voltage waveform shown in Figure 9.10 is shown in Figure 9.12.

From Figure 9.12, the maximum current THD is 9% and it is observed in R phase. The individual harmonic order 5th is predominant (about 5%) among other harmonic orders.

The voltage and current waveform of 30 kW charger at 16:28:59 hours when it is charging the e-auto 2 with battery SoC is 45% and e-auto 3 with battery SoC is 83% is shown in Figure 9.13.

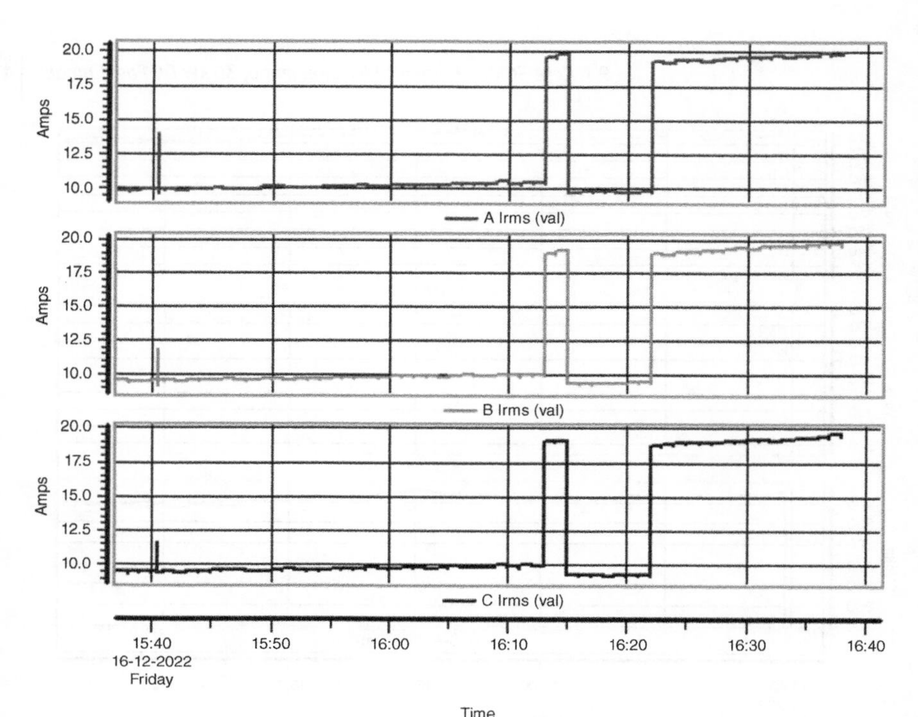

	Min	Max	Avg
AIVrms	9.607	20.04	12.65
BIVrms	9.104	19.75	12.23
CIVrms	9.151	19.61	12.16

Figure 9.6 Current RMS trend during the measurement period.

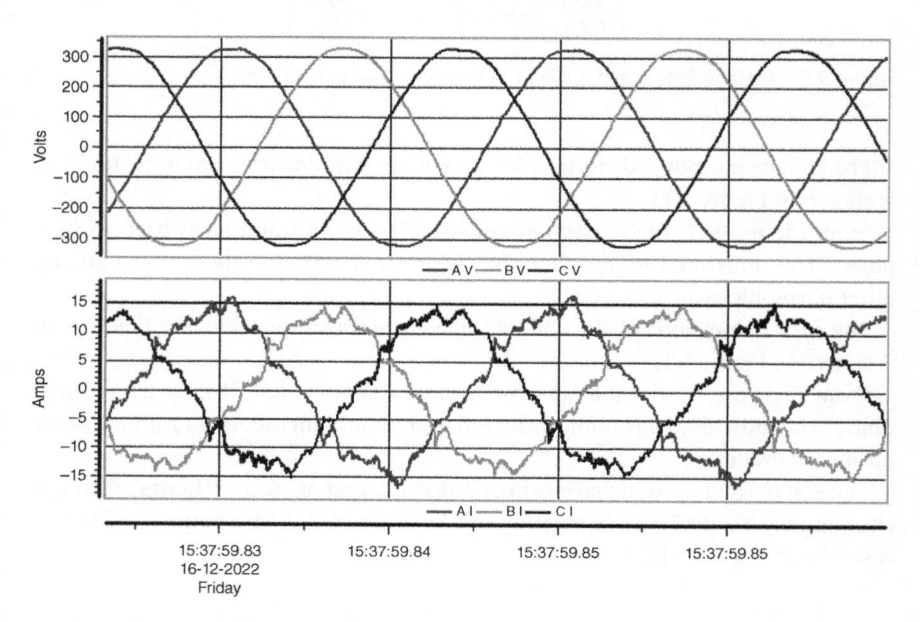

Figure 9.7 Voltage and current waveform of 30 kW charger.

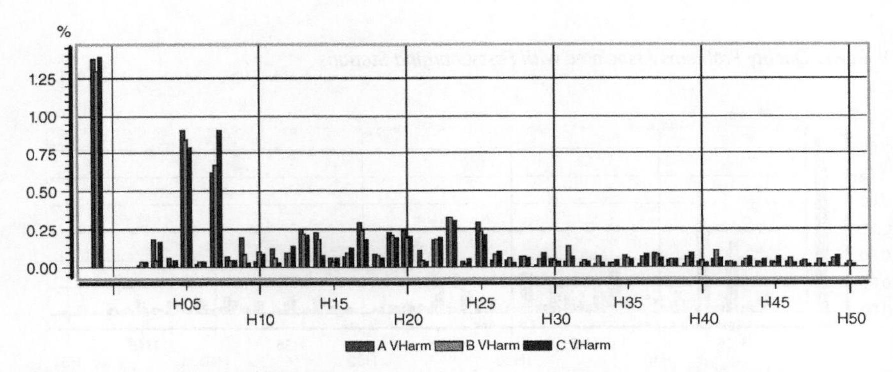

Figure 9.8 Voltage harmonic distortion for the voltage waveform shown in Figure 9.7.

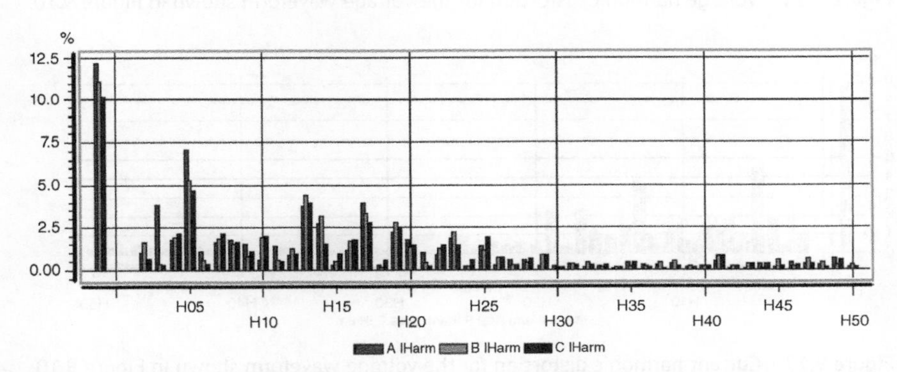

Figure 9.9 Current harmonic distortion for the voltage waveform shown in Figure 9.7.

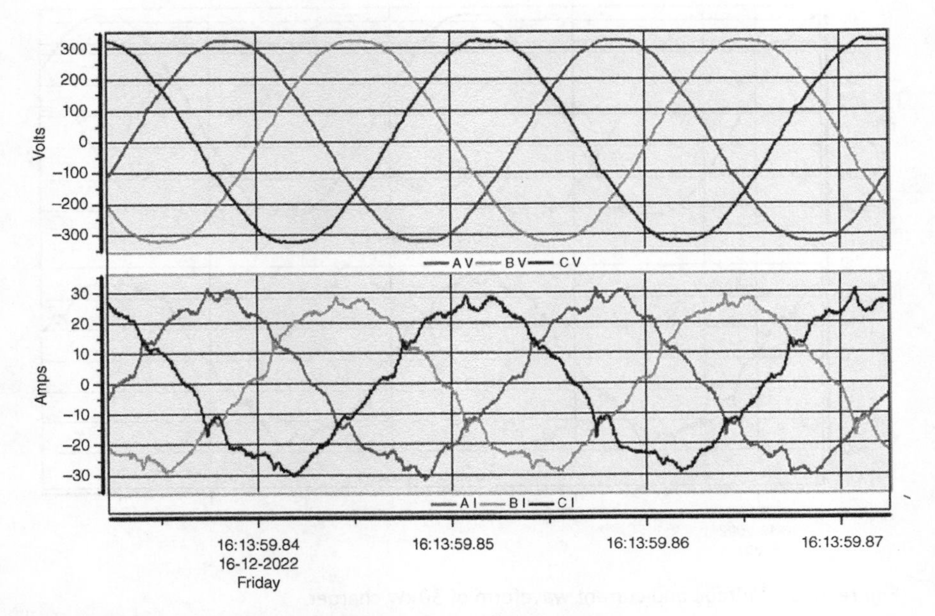

Figure 9.10 Voltage and current waveform of 30 kW charger.

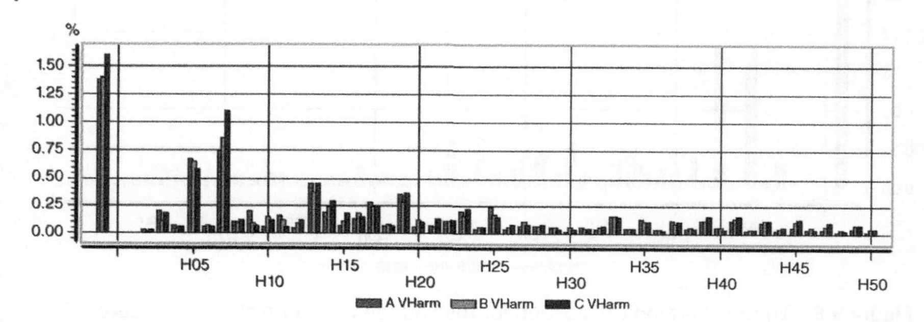

Figure 9.11 Voltage harmonic distortion for the voltage waveform shown in Figure 9.10.

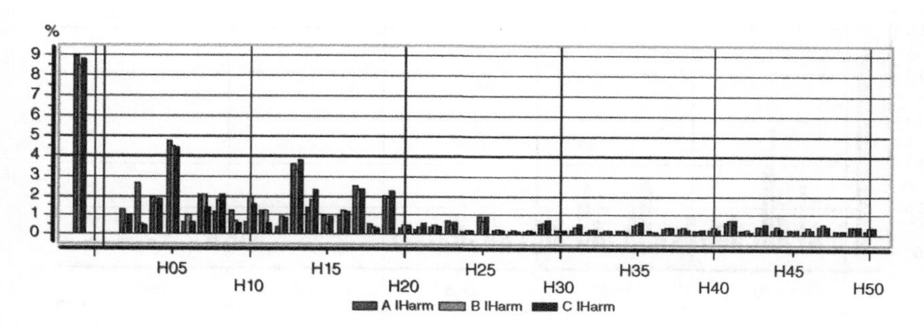

Figure 9.12 Current harmonic distortion for the voltage waveform shown in Figure 9.10.

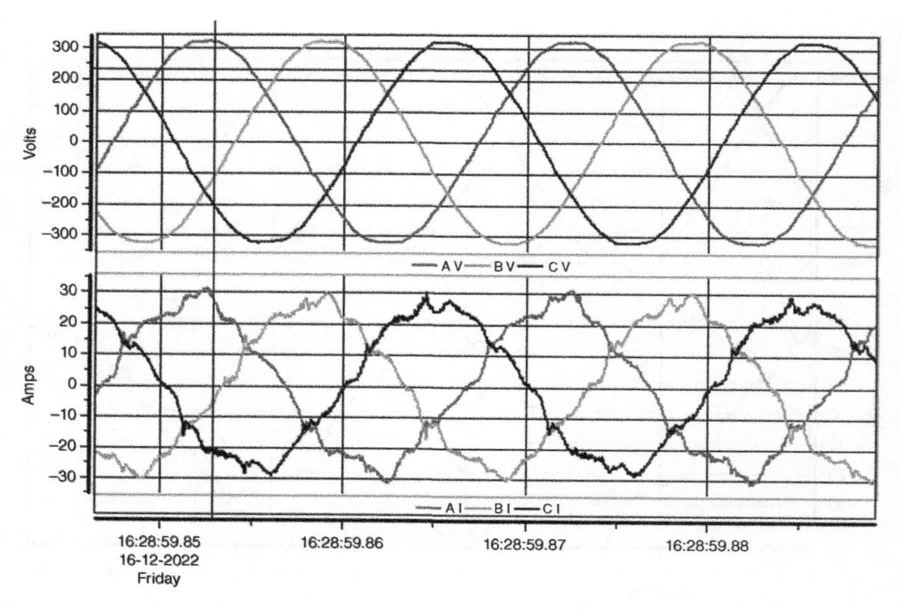

Figure 9.13 Voltage and current waveform of 30 kW charger.

The voltage harmonic distortion for the voltage waveform shown in Figure 9.13 is shown in Figure 9.14.

From Figure 9.14, the maximum voltage THD is 1.35% and it is observed in Y phase. The individual harmonic order 5th and 7th are predominant (about 0.8%) among other harmonic orders.

The current harmonic distortion for the voltage waveform shown in Figure 9.13 is shown in Figure 9.15.

From Figure 9.15, the maximum current THD is 10.5% and it is observed in R phase. The individual harmonic order 5th is predominant (about 8.5%) among other harmonic orders.

The voltage THD trend for the entire monitoring duration is shown in Figure 9.16 and current THD trend is shown in Figure 9.17.

From Figure 9.16, the maximum voltage THD in R phase is 1.69%, Y phase is 1.61%, and B phase is 1.63%.

From Figure 9.17, the maximum current THD in R phase is 13%, Y phase is 10.52%, and B phase is 10.44%.

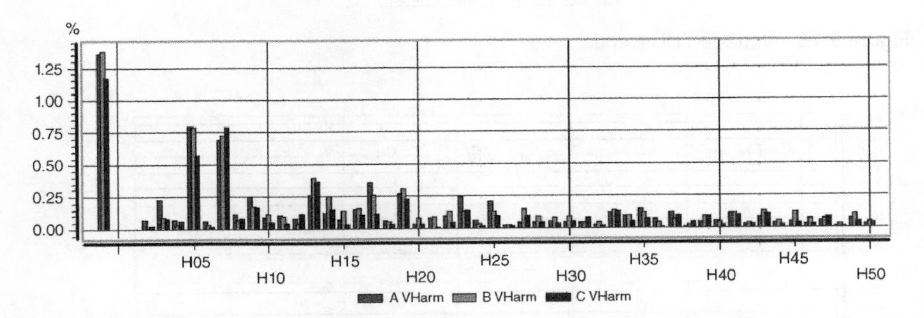

Figure 9.14 Voltage harmonic distortion for the voltage waveform shown in Figure 9.13.

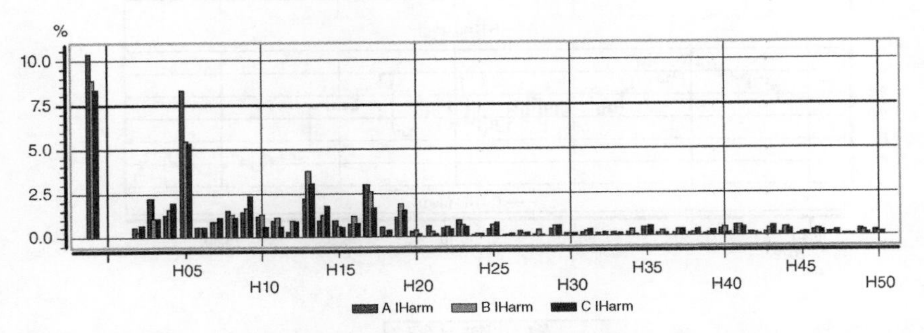

Figure 9.15 Current harmonic distortion for the voltage waveform shown in Figure 9.13.

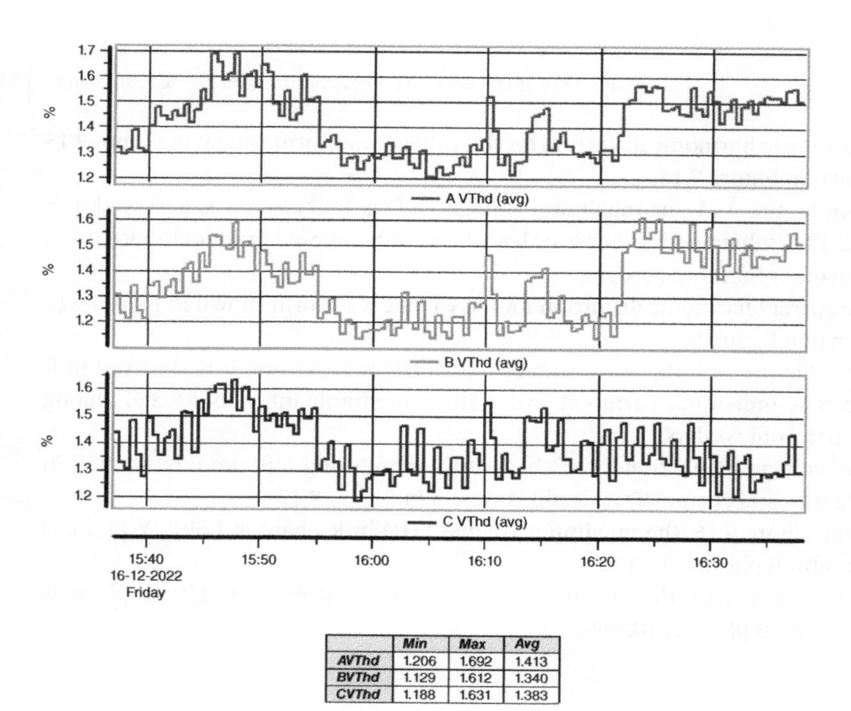

	Min	Max	Avg
AVThd	1.206	1.692	1.413
BVThd	1.129	1.612	1.340
CVThd	1.188	1.631	1.383

Figure 9.16 Voltage THD trend.

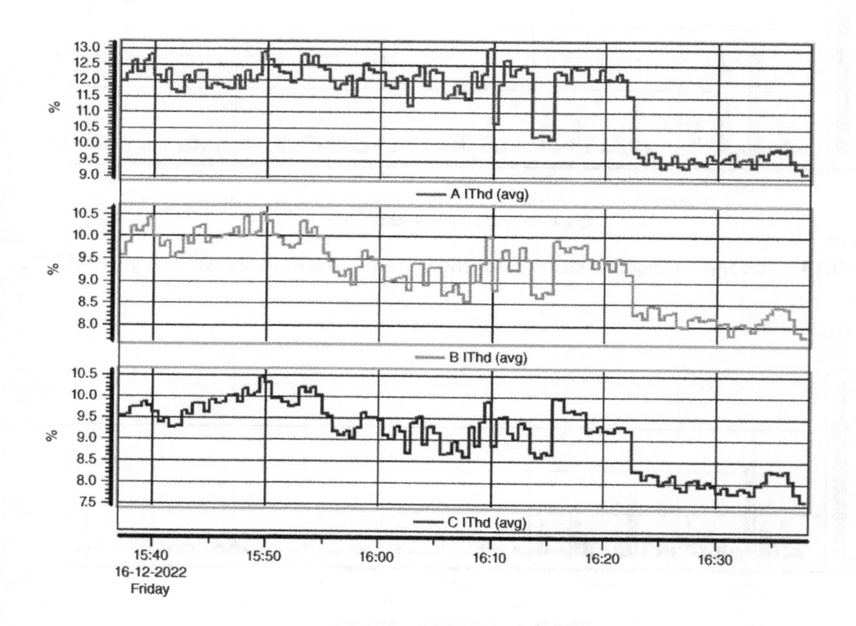

	Min	Max	Avg
AIThd	9.125	13.02	11.43
BIThd	7.777	10.52	9.220
CIThd	7.583	10.44	9.102

Figure 9.17 Current THD trend.

9.8 Conclusion

PQ is one of the problems in the power system which gets more attention in the past two decades. In a fast-charging station, the poor PQ has an impact in two ways, one is the grid on utility equipment and the second is the impact on fast-charging station equipment. The PQ characteristics like voltage sag, voltage unbalance, etc., from the utility grid affects the performance of fast chargers and harmonics, flicker, etc., created by the fast charger have an impact on the utility equipment and their performance. This chapter discusses the PQ problems associated with the fast-charging station and a case study for harmonic measurement of 30 kW DC fast charger.

References

1 IEEE Std 2030.1.1-2015 (2015). *IEEE Standard Technical Specifications of a DC Quick Charger for Use with Electric Vehicles*. IEEE Vehicular Technology Society.

2 Sivaraman, P. and Sharmeela, C. (2021). Power quality problems associated with electric vehicle charging infrastructure. In: *Power Quality in Modern Power Systems* (ed. P. Sanjeevikumar, C. Sharmeela, and J.B. Holm-Nielsen). London: Academic Press. https://doi.org/10.1016/B978-0-12-823346-7.00005-0.

3 Wang, H., Song, Q., Zhang, L., Wen, F., and Huang, J. (2012). Load characteristics of electric vehicles in charging and discharging states and impacts on distribution systems. *Proceedings of International Conference on Sustainable Power Generation and Supply,* Hangzhou, China (8–9 September 2012). https://doi.org/doi:10.1049/cp.2012.1837.

4 IEC 61000-4-30-2015 (2015). *Testing and Measurement Techniques – Power Quality Measurement Methods*. IEC.

5 IEEE 1100-2005 (2005). *IEEE Recommended Practice for Powering and Grounding Electronic Equipment*. IEEE.

6 IEEE Std 1159-2019 (2019). *IEEE Recommended Practice for Monitoring Electric Power Quality*. IEEE.

7 Sivaraman, P. and Sharmeela, C. (2021). Power quality characteristics. In: *Power Quality in Modern Power Systems* (ed. P. Sanjeevikumar, C. Sharmeela, and J.B. Holm-Nielsen). London: Academic Press. https://doi.org/10.1016/B978-0-12-823346-7.00005-0.

8 Wang, Z., Guo, X., Li, J., and Wang, X. (2016). Impact of voltage sags on electric-vehicle charger and critical voltage sag determination. *IEEE Transactions on Power Delivery* 31 (3): 1397–1399. https://doi.org/10.1109/TPWRD.2015.2473100.

9 Pillai, J.R. and Bak-Jensen, B. (2010). Impacts of electric vehicle loads on power distribution systems. *IEEE Vehicle Power and Propulsion Conference 2010,* Lille, France.

10 Society of Automotive Engineers Standard J2894–1 (2011). Power Quality Requirements for Plug-In Electric Vehicle Chargers.

11 Letha, S.S. and Bollen, M. (2021). Technical Report on Impact of Electric Vehicle Charging on the Power Grid, Lulea University of Technology, Sweden, pp. 1–44. https://www.diva-portal.org/smash/get/diva2:1530550/FULLTEXT02.pdf (accessed 15 March 2023). ISBN 978-91-7790-763-3.

12 Yaghoobi, J., Alduraibi, A., Martin, D. et al. (2020). Impact of high-frequency harmonics (0–9 kHz) generated by grid-connected inverters on distribution transformers. *International Journal of Electrical Power and Energy Systems* 122 (April): 106177.

13 Slangen, T.M.H., van Wijk, T., Ćuk, V., and Cobben, J.F.G. (2020). The propagation and interaction of Supraharmonics from electric vehicle chargers in a low-voltage grid. *Energies* 13 (15): 3865. https://doi.org/10.3390/en13153865.

14 Lucas, A., Bonavitacola, F., Kotsakis, E., and Fulli, G. (2015). Grid harmonic impact of multiple electric vehicle fast charging. *Electric Power Systems Research* 127: 13–21. https://doi.org/10.1016/j.epsr.2015.05.012.

15 IEEE Std 3002.8-2018 (2018). *Recommended Practice for Conducting Harmonic Studies Ana Analysis of Industrial and Commercial Power Systems*. IEEE.

16 IEEE Std 519-2022 (2022). *IEEE Standard for Harmonic Control in Electric Power Systems*. IEEE.

17 Sivaraman, P. and Sharmeela, C. (2021). Power system harmonics. In: *Power Quality in Modern Power Systems* (ed. P. Sanjeevikumar, C. Sharmeela, and J.B. Holm-Nielsen). London: Academic Press. https://doi.org/10.1016/B978-0-12-823346-7.00005-0.

18 Slangen, T.M.H., van Wijk, T., Ćuk, V., and Cobben, J.F.G. (2020). The harmonic and supraharmonic emission of battery electric vehicles in The Netherlands. In: *3rd International Conference on Smart Energy Systems and Technologies, SEST 2020*, Istanbul, Turkey (September 2020), 1–6. https://doi.org/10.1109/SEST48500.2020.9203533.

19 IEC 61000-4-30 (draft)-2022 (2022). *Testing and Measurement Techniques – Power Quality Measurement Methods*. IEC.

20 Sivaraman, P. and Sharmeela, C. (2020). Introduction to Electric Distribution System. *Advances in Computer and Electrical Engineering* 1–31. https://doi.org/10.4018/978-1-7998-1230-2.ch001.

21 Sivaraman, P. and Sharmeela, C. (2020). Existing Issues Associated With Electric Distribution System. *Advances in Computer and Electrical Engineering* 32–47. https://doi.org/10.4018/978-1-7998-1230-2.ch002.

10

Standards for Fast-Charging Infrastructure

10.1 Introduction

The standards are used to provide the requirements of the particular equipment or EV ecosystem as a whole to meet the design, construction, operation, performance, testing, safety, interoperability, etc., when they are operating. There are different regulatory or governing bodies available across the globe and they are developing the standards for the manufacturing and operation of electric vehicles (EVs) including their charging requirements. The international standards organization like the International Organization for Standards (ISO), International Electrotechnical Commission (IEC), Institute of Electrical and Electronics Engineers (IEEE), and Society of Automotive Engineers (SAE), etc., and national standards organizations like the Bureau of Indian standards, British Standards Institute, etc., are developing the standards in the entire EV ecosystem including charging stations. The objective of the standards development for EV charging infrastructure is design, construction, function, operation, testing, communication, safety, interoperability, etc., to meet their requirements for reliable operation.

Generally, the standards for EV charging infrastructure consist of charger standards, charging station standards, cables and connector standards, communication standards, and safety standards including switchgear, grounding, and lightning protection.

Based on the type of power transfer to EV, EV charging is classified into two types:

- Conductive charging or wired charging
- Inductive charging or wireless charging

Fast-Charging Infrastructure for Electric and Hybrid Electric Vehicles: Methods for Large-Scale Penetration into Electric Distribution Networks, First Edition. Sivaraman Palanisamy, Sharmeela Chenniappan, and Sanjeevikumar Padmanaban.

Published 2023 by John Wiley & Sons, Inc.

Based on the type of electricity used to charge the EVs, EV charging is further classified into:

- AC charging
- DC charging

The standard and their requirements for each type of charging are different. For example, IEC 61851 and SAE J1772 are the standards used for wired charging, in the meantime IEC 61980 and SAE J2954 are the standards used for wireless charging.

10.2 IEC Standards

10.2.1 IEC 61851

The IEC 61851 covers the electric vehicle supply equipment used to charge EVs including light EVs. The broad scope of this IEC 61851 includes electrical, mechanical, electro magnetic compatibility (EMC), performance, and communications requirements for EVSE. IEC 61581 covers wired or conductive AC and DC charging for ON-board as well as OFF-board EVSE with a rated voltage of up to 1000 V AC or up to 1500 V DC and rated output voltage of up to 1000 V AC or up to 1500 V DC to EVs [1].

The features covered in this standard are:

- EVSE operating conditions and characteristics.
- Connection between the EV and EVSE and their specification.
- Electrical safety requirements for the EVSE.

As per IEC 61851, EVSE is classified as Mode 1, Mode 2, Mode 3, and Mode 4 based on charging modes. The features of these charging modes are listed in Table 10.1. Charging modes 1–3 are used for AC charging and mode 4 is used for DC fast charging.

- Mode 1 is a simple connection established for charging between the EV and AC socket outlet. The controlling pilot pin is not present in this type. Also, it provides an additional protective earth conductor for safety. This charging mode is not allowed or prohibited in a few countries.
- Mode 2 is a connection established between an EV and an AC socket with a control pilot pin. It has personal protection from electric shocks between the EV and the standard plug. It is limited in use in some countries based on values of current, voltage, and location.

Table 10.1 Features of charging modes as per IEC 61851. *Source:* Adapted from [1].

			IEC 61851 standard			
S. No.	**Parameter**		**Mode 1**	**Mode 2**	**Mode 3**	**Mode 4**
1	Voltage (V)	Single phase	250	250	—	DC 200-600
		Three phase	480	480	480	
2	Current (A)		16	32	250	400
3	Power (kW)		13.3	26.6	184.4	240
4	Connector		Household	Household	IEC 62196 Type 1–3	IEC 62196 Type 2

- Mode 3 is a connection establishment between EV and AC EVSE connected to the supply system with a control pilot function for controlled charging. It has a protective earthing conductor for the EV socket or vehicle connector.
- Mode 4 is a connection established between EV to DC EVSE with a control pilot function. It also has a protective conductor connected to the vehicle connector.

This standard (i.e. IEC 61851) also provides communication requirements between EV and EVSE. The communication between EV and EVSE is optional for Mode 1, Mode 2, and Mode 3 connections. Mode 4 connection between EV and EVSE is required to have communication as per IEC 61851-24. This communication allows the control of EVSE by EV.

IEC 61851 listed some mandatory functions of Mode 2, 3, and 4 related to EVSE operation:

- Continuous protective conductor checking.
- Verification of proper connection between EVSE and EV.
- Energization and de-energization of the power supply.
- Maintaining current within maximum limits.

IEC 61851 is split into several subparts as listed below:

- IEC 61851 – Part 1: General Requirements. This part provides the general requirements that provide as a foundation for all the subsequent standards in the series.
- IEC 61851 – Part 21-1: Electric vehicle onboard charger EMC requirements for conductive connection to an AC/DC supply. This part of the standard provides the requirements for EMC onboard the vehicle [3].

- IEC 61851 – Part 21-2: EMC requirements for OFF-board electric vehicle charging systems. This part of the standard provides all requirements for AC and DC EV supply equipment [4].
- IEC 61851 – Part 23: DC electric vehicle charging station. This part of the standard provides the requirements for DC charging stations both permanently wired and cable and plug connected [5].
- IEC 61851 – Part 24: Digital communication between a DC EV charging station and an electric vehicle for control of DC charging. This part of the standard covers the requirements for communication between the DC charging stations and EV as per IEC 61851 – Part 23 [6].

IEC 61851-3 subseries is presently in the development stage and is intended to cover EV supply equipment with a DC output not exceeding 120V. The IEC 61851-3 subseries are as follows [1]:

- Part 3-1: Electric vehicles conductive power supply system – Part 3-1: General Requirements for Light Electric Vehicles (LEV) AC and DC conductive power supply systems.
- Part 3-2: Electric vehicles conductive power supply system – Part 3-2: Requirements for Light Electric Vehicles (LEV) DC off-board conductive power supply systems.
- Part 3-3: Electric vehicles conductive power supply system – Part 3-3: Requirements for Light Electric Vehicles (LEV) battery swap systems.
- Part 3-4: Electric vehicles conductive power supply system – Part 3-4: Requirements for Light Electric Vehicles (LEV) communication.
- Part 3-5: Electric vehicles conductive power supply system – Part 3-5: Requirements for Light Electric Vehicles communication – predefined communication parameters.
- Part 3-6: Electric vehicles conductive power supply system – Part 3-6: Requirements for Light Electric Vehicles communication – Voltage converter unit.
- Part 3-7: Electric vehicles conductive power supply system – Part 3-7: Requirements for Light Electric Vehicles communication – Battery system.

10.2.2 IEC 61980 Electric Vehicle Wireless Power Transfer Systems

The IEC 61980 covers the wireless methods for charging the EVs on the road and the supply voltage is up to 1000V AC and up to 1500V DC [7]. The aspects provided in this standard are:

- Characteristics and operating conditions of the supply device.
- Electrical safety required for a supply device.

- Communication between vehicle and supply device.
- Supply device EMC requirements.
- Efficiency, alignment, and other activities to enable wireless power transfer (WPT).

As per IEC 61980, the supply device is classified into the following:

- Power transfer technology: The magnetic field is used as a power transfer technology as specified in IEC 61980-2 and IEC 61980-3.
- Environment condition: Based on the environmental condition, the use of supply devices is classified into indoor, outdoor, and industrial locations exposed to pollution.
- Electric connection to supply network: Based on the electric connection to supply network or system, supply devices are classified into permanently connected, plug and cable connected.
- Protection from electric shock: Based on the protection against electric shock, the supply devices are classified into class I equipment and class II equipment.
- Access: Based on the access to the location of installation, supply devices are classified into equipment for locations with restricted access and without access.

The communication between the EV and supply is important for starting, controlling, and terminating the WPT. The requirements for communication between EV and supply device shall be as per the technology-specific parts of the IEC 61980 series standards.

The requirements of harmonics to be included in the WPT system are expected to connect with a public supply network (i.e. electric distribution system) and the scope is IEC 61000-3-2 or IEC 61000-3-12 [7].

- The supply device of WPT has an input current of up to 16 A per phase and is connected to the low-voltage AC public distribution system, the harmonic requirements have to comply with IEC 61000-3-2.
- The supply device of WPT has an input current exceeding 16 A per phase and less than or equal to 75 A and is connected to the low-voltage AC public distribution system, the harmonic requirements have to comply with IEC 61000-3-12.

The disturbance is caused by voltage fluctuation and flicker due to the WPT system and it is expected to connect with a public supply network (i.e. electric distribution system):

- The supply device of WPT has an input current of up to 16 A per phase, the IEC 61000-3-3 compliance is required.

- The supply device of WPT has an input current exceeding 16 A per phase and less than or equal to 75 A and is connected to the low-voltage AC public distribution system, the IEC 61000-3-11 compliance is required.

The IEC 61980 series was published in multiple subparts as listed below:

- IEC 61980-1 provides the general requirements for WPT systems for EVs.
- IEC 61980-2 specifically provides the requirements for magnetic field wireless power transfer (MF-WPT) for EVs. Also, it provides the specific requirements for system activities and communication required between the electric road vehicle side and the off-board side [8].
- IEC 61980-3 specifically provides the power transfer requirements for the off-board side of MF-WPT systems for electric road vehicles.

10.2.3 IEC 62196 Plugs, Socket-Outlets, Vehicle Connectors, and Vehicle Inlets – Conductive Charging of Electric Vehicles

This standard provides the requirements for plugs, socket outlets, vehicle connectors, vehicle inlets, and cable assemblies as per IEC 61851 [9]. It covers the electrical, mechanical, and performance requirements for plugs, vehicle connectors, and vehicle inlets connected between EV and EVSE.

This standard shall be used for:

- 690 V, 50 Hz, or 60 Hz AC supply and a rated current not more than 250 A.
- 1500 V DC supply and rated current not more than 800 A.
- Ambient temperature in the range between −30 °C and +40 °C.

This standard provides the rated voltage and current values for signal control purposes. Also, details related to protection against electrical shocks, tests circuit and its procedure, interlocks, degree of protection (IP degree), normal operation, and temperature rise are also covered in the standard series.

The IEC 62196 series of standards consists of the following subparts:

Part 1: General requirements, comprising clauses of a general character.

Part 2: Dimensional compatibility and interchangeability requirements for AC pin and contact-tube accessories [10].

Part 3: Dimensional compatibility and interchangeability requirements for DC and AC/DC pin and contact-tube vehicle couplers [11].

Part 3-1: Vehicle connector, vehicle inlet, and cable assembly intended to be used with a thermal management system for DC charging.

Part 4: Dimensional compatibility and interchangeability requirements for DC pin and contact-tube accessories for Class II or Class III applications.

Part 6: Dimensional compatibility and interchangeability requirements for DC pin and contact-tube couplers for applications using a system of protective electrical separation.

10.2.4 IEC TR 62933-2-200 Electrical Energy Storage (EES) Systems – Part 2-200: Unit Parameters and Testing Methods – Case Study of EES Systems Located in EV Charging Station with PV

This standard presents the case study for EES systems installed in EV charging stations with a combination of solar PV systems (ESS-PV-EV charging stations) connected to a voltage level of 20 kV or lesser. The EES is an essential option for making the fast-charging stations more grid friendly, improving the utilization of solar PV systems, and increasing the financial income of charging stations [12].

From the case studies presented in this standard, EES systems show exceptional performance when they are operating in different modes such as power smoothing, peak shaving, time-of-use (ToU) price arbitrage, load tracing, and ancillary services to the grid.

This standard includes the following features or elements in it:

- Overview of general PV-EES-EV charging stations.
- Operational analysis of EES systems in typical project cases.
- Summary and recommendation of EES systems' operation modes.

10.2.5 IEC 62893 Charging Cables for Electric Vehicles for Rated Voltages up to and Including 0.6/1 kV

This standard (i.e. IEC 62893) provides the requirements of cable between EV and EVSE for flexible applications under harsh environmental conditions such as dimensions, construction, and test requirements for cables with extruded insulation and sheath for a rated voltage up to and including 0.6/1 kV AC or up to and including 1500 V DC [13]. These cables are intended to supply the power and the same time communication between an EV/Plug-in Hybrid EV (PHEV) and EVSE. The maximum conductor temperature for the cables in this part of IEC 62893 is 90 °C. The IEC 62893-3 provides the requirements of charging cables for AC charging (modes 1–3) and IEC 62893-4 for DC charging (mode 4).

The IEC 62893-4 further has two parts:

- IEC 62893-4-1 Charging cables for electric vehicles for rated voltages up to and including 0.6/1 kV – Part 4-1 Cables for DC charging according to mode 4 of IEC 61851-1 – DC charging without use of a thermal management system.

- IEC 62893-4-2 Charging cables for electric vehicles for rated voltages up to and including 0.6/1 kV – Part 4-2 Cables for DC charging according to mode 4 of IEC 61851-1 – Cables intended to be used with a thermal management system.

The charging cables used in fast-charging stations need to comply with either IEC 62893-4-1 or IEC 62893-4-2.

IEC 62893-4-1 gives the technical requirements of the cables for mode 4 DC charging without the use of a thermal management system as follows [14]:

Rated voltage: 1.5 kV DC
Power cores: 4–150 mm^2 for two or more cores
Auxiliary power cores (optional): 2.5–6 mm^2 – two cores
Temperature sensor cores (optional): Minimum size is 0.5 mm^2
Conductor: Plain or tinned annealed copper comply with Class 5 of IEC 60228
Number of cores: Two-core cable and three-core cable
The preferred color for DC cables:

- Two-core cable: No preferred color requirements
- Three-core cable: Green and yellow, no preferred color for other cores

Insulation material:

- Power cores: Compound of EVI-2
- Pilot, control, and any other additional cores: EVI-1 or EVI-2

Metallic screen (optional): Braid of plain or tinned copper wire
Sheath: Compound of EVM-1, EVM-2, or EVM-3

The recommended use of cables without thermal management for EV charging is listed in Table 10.2.

IEC 62893-4-2 gives the technical requirements of the cables for mode 4 DC charging with a thermal management system as follows [15]:

Rated voltage: 1.5 kV DC
Power cores: 16–150 mm^2 for two or more cores

Table 10.2 Recommended use of cables without thermal management for EV charging. *Source:* Adapted from [14].

S. No.	Code designation	Cable type	Recommended use	Application	Remarks
1	62893 IEC 126	EVM-1	Mode 4 as per IEC 61851-1	Indoor and outdoor	—
2	62893 IEC 127	EVM-2			
3	62893 IEC 128	EVM-3		Outdoor	Used only for outdoor applications because it will release the halogen content during fire

Auxiliary power cores (optional): 2.5–6 mm^2 – two cores
Temperature sensor cores (optional): Minimum size is 0.5 mm^2
Conductor: Plain or tinned annealed copper comply with Class 5 of IEC 60228.
Number of cores: Two-core cable and three-core cable
The preferred color for DC cables:

- Two-core cable: No preferred color requirements
- Three-core cable: Green and yellow, no preferred color for other cores

Insulation material:

- Power cores: Compound of EVI-2
- Pilot, control, temperature sensor, and auxiliary power cores: EVI-1 or EVI-2

Metallic screen (optional): Braid copper with minimum 80% optical coverage
Tubes: Fluid-filled tubes should be made resistant to the media used as a coolant. It has to withstand aging at 120 °C temperature for 168 hours in dry air. If the tube is also used as an insulator, it has to comply with the requirements of EVI-2.
Sheath: Compound of EVM-1, EVM-2, or EVM-3
Liquid coolants: The coolant type used in the cable shall be specified by the manufacturer of the system. It has to satisfy the test requirements of IEC 62893-4-2.

The recommended use of cables with thermal management for EV charging is listed in Table 10.3.

Table 10.3 Recommended use of cables with thermal management for EV charging. *Source:* Adapted from [15].

S. No.	Code designation	Cable type	Recommended use	Application	Limitations	Remarks
1	62893 IEC 129	EVM-1	Mode 4 as per IEC 61851-1 with the thermal management system	Indoor and outdoor	During normal operating conditions, the cable's maximum surface voltage should not exceed 60 °C	—
2	62893 IEC 130	EVM-2				
3	62893 IEC 131	EVM-3	Mode 4 as per IEC 61851-1 with the thermal management system	Outdoor		Used only for outdoor applications because it will release the halogen content during fire

10.2.6 IEC 60364-7-722 Low-Voltage Electrical Installations – Part 7-722: Requirements for Special Installations or Locations – Supplies for Electric Vehicles

This standard provides the requirements for distribution circuits at the connecting point:

- Distribution circuits intended to supply energy to EVs
- Distribution circuits feeding back the electricity to the grid from EVs

The requirement specified in this standard is applicable to IEC 61851 (all parts) for conducting EV charging stations and WPT as per IEC 61980 (all parts). This standard provides the requirements for [16]:

- Protection against electric shock
- Protective measure: automatic disconnection of supply
- Protective measure: electrical separation
- Protection against voltage disturbances and electromagnetic disturbances
- Protection against voltage and electromagnetic disturbances
- Protection against transient overvoltages of atmospheric origin or due to switching
- Measures against electromagnetic influences
- Selection and erection of electrical equipment
- Devices for protection against indirect contact by automatic disconnection of supply
- Devices for protection against overcurrent
- Earthing arrangements and protective conductors

10.3 IEEE Standards

10.3.1 IEEE Std 2030.1.1-2021 IEEE Standard for Technical Specifications for a DC Quick and Bidirectional Charger for Use with Electric Vehicles

This standard provides the requirements for the EVs and DC quick charger for efficient and rapid charging of EVs.

The general requirements for a DC quick charger are listed below [17]:

- DC quick charger is able to indicate to the user the status of the charging process and it has to take any corrective actions if essential.
- During the entire charging period, both the DC quick charger and the vehicle shall be confirmed they are physically and/or electrically connected.

- The DC quick charger shall have the feature to stop the charging in the event of communication between the vehicle and the charger gets interrupted (via the communications interface).
- Whenever the protective conductor between the vehicle and the charger gets disconnected, the DC quick charger has to stop charging within 10 seconds time period.
- The DC quick charger shall have the feature to detect or identify the short-circuit, ground faults, and loss of isolation.
- The DC quick charger shall have the feature of protection against overvoltage.
- Whenever the connector is not connected to the vehicle, the DC quick charger shall be designed in such a way that, it will not provide the level of voltage that is dangerous or harmful to the human body.
- The system shall be designed in such a way that it will not allow the users to touch the electrically energized parts of the charger and vehicle.
- The DC quick charger shall have the feature of detecting earth leakage and automatically disconnecting the circuit and preventing the users from electric shock hazards.
- The DC quick charger or charging connector shall be equipped with measures (e.g. the plastic cap on connector power terminals) to reduce the risk of contact with exposed live parts as a measure against the remaining electric charge on the charging connector.
- The DC quick charger shall have the feature of protection against overload and short-circuits in the AC circuits or internal circuits.
- The charging system shall be designed in such a way that the voltage level between any accessible conductive parts, including the charging cable and charging connector, and any grounding parts is reduced to less than 60 V within 1 seconds or less, after the connector is removed from station or vehicle.

The communication requirements are listed below:

- The vehicle and charger shall exchange the parameters or data through the communication interface.

The indoor/outdoor requirements for conducting charging:

- For the outdoor application, the IP degree must be more than or equal to IP44.
- For the indoor application, IP degree must be more than or equal to IP41.
- Ambient temperature shall be −10 °C to 40 °C.
- Ambient humidity shall be 5–95%.
- The DC quick charger shall be suitable for work at an altitude of 1000 m or lower.

Other requirements are provided in Annex A and Annex B of IEEE Std 2030.1.1-2021.

10.3.2 IEEE Std 2836-2021 IEEE Recommended Practice for Performance Testing of Electrical Energy Storage (ESS) System in Electric Charging Stations in Combination with Photovoltaic (PV)

This standard defines performance test requirements for EES systems for complex applications such as smoothing the output power from the solar PV system and regulating the electric vehicle load. The requirement for such standards is growing due to the technological developments in the manufacturing of converters and energy storage modules, and the increased usage and application of energy storage systems in charging stations [18].

In order to verify the compliance of the system with respect to the requirements, standardized test procedures are essential for this purpose. This IEEE-recommended practice provides the performance tests for EES systems and their documentation; based on the engineering practice, it provides references to duty cycle-based test protocols.

From the results, users can be able to evaluate the application of the EES system in electric vehicle charging infrastructure or stations in combination with solar PV systems.

10.4 SAE Standards

10.4.1 SAE J1772 SAE Electric Vehicle and Plug-in Hybrid Electric Vehicle Conductive Charge Coupler

This standard provides the general requirements for electrical, physical, functional, and performance to establish the wired or conductive charging of EVs/PHEVs in North America. It provides the common EV/PHEV and EVSE conductive charging method including functional, operational, and dimensional requirements for the mating connector as well as the vehicle inlet [19].

The SAE J1772-2017 defines the EVs charging into four levels, namely AC level 1, AC level 2, DC level 1, and DC level 2. The technical details of AC level 1, AC level 2, DC level 1, and DC level 2 are listed in Table 10.4.

10.4.2 SAE J2894-1 2019 Power Quality Requirements for Plug-In Electric Vehicle Chargers

This standard provides the recommended practice for Plug-In EV chargers including both ON-board as well as OFF-board regarding power quality. This standard enables the manufacturers of vehicles as well as charging equipment, electric utilities of DISCOMs, and others who are involved in EV ecosystem and concern about power quality to make a reasonable decision on design aspects [20].

Table 10.4 Technical details of AC level 1, AC level 2, DC level 1, and DC level 2.
Source: Adapted from [19].

S. No.	Charge method	Voltage (V)	Maximum current (A)	Maximum power (kW)
1	AC Level 1 – single phase	120	12 or 16	1.92
2	AC Level 2 – single phase	208 or 240	24–80	19.2
3	DC Level 1	50–1000	80	80
4	DC Level 2	50–1000	400	400

This standard focuses on three main aspects:

- Identifying the parameters of the Plug-In EV charger that are essential to be controlled in order to maintain the power quality of the AC service.
- Identifying the characteristics of AC service that may significantly impact the charger performance.

To identify values for power quality, susceptibility, and power control parameters which are based on the current US and international standards. Implementation of these values should be technically feasible for the charger and cost-effective.

Some of the requirements are [2]:

- During the voltage sag, EV chargers have to remain energized for voltage reduced to 80% of nominal voltage for up to two seconds.
- EV chargers shall have fault ride-through capability to ride through the complete loss of voltage during a fault for up to 12 seconds.

10.5 ISO 17409 Electrically Propelled Road Vehicles – Connection to an External Electric Power Supply – Safety Requirements

This standard provides the electrical safety requirements for wired connections of EVs to an external electric power supply using a plug or vehicle inlet. The requirements specified in this standard is applicable only to the circuits delivering the power supply to vehicles and dedicated power supply control functions between EVs connection to an external power supply [20]. This standard provides the requirements for:

- Protection under single failure conditions
- Isolation resistance
- Protection from unintended voltage

- Insulation coordination
- Overload protection
- Short-circuit protection for AC connection and DC connection
- Arc protection for DC connections
- Requirements for the plug and cable
- Requirements for the vehicle inlet
- Operational requirements
- Test procedures

10.6 CEA Technical Standards in India

10.6.1 Technical Standards for Connectivity of the Distributed Generation Resources – February 2019

The standards for a charging station, prosumers, or a person connected or seeking connectivity to the electricity system are as follows [21]:

- The charging station shall have a reliable protection system to identify the various faults and abnormal conditions, and isolate the faulted equipment or line automatically.
- DISCOM shall carry out the adequacy and stability studies before connecting the charging station to the power system.
- The current harmonic injection from the charging station shall not exceed the limits specified by IEEE Std 519-2022 at the point of common coupling.
- The measuring and metering of harmonics shall be a continuous process with power quality meters complying with the provisions of IEC 61000-4-30 Class A.
- The charging station seeking connectivity at a voltage level of 11 kV or above shall install power quality meters.

10.6.2 Technical Standards for Measures Relating to Safety and Electric Supply – June 2019

The guidelines for safety provisions for electric vehicle charging stations are as follows [22]:

- Electric vehicle charging stations shall have overload protection on input supply and output supply fittings.
- The EV charging station shall have a protection against the uncontrolled reverse power flow from vehicle.
- All EVSE points shall be installed at least 800 mm above from the finished ground level.

- A cord extension set shall not be used in addition to the supply lead for establishing the connection of the EVSE and EV.
- Any adaptors shall not be used between the vehicle connector to a vehicle inlet.
- The distance between the charging point (EVSE) and EV (parked in a place) shall be within 5 m.
- Portable socket outlets are not allowed.
- The charging station shall be protected from lightning by means of suitable lightning protection system as per Indian Standards (IS/IEC 62305).
- One second after having disconnected the electric vehicle from the supply (mains), the voltage between accessible conductive parts or any accessible conductive part and earth shall be less than or equal to 42.4 V peak (30 V rms), or 60 V DC, and the stored energy available shall be less than 20 J (as per IEC 60950) and if the voltage is greater than 42.4 V peak (30 V rms) or 60 V DC, or the energy is 20 J or more, a warning label shall be attached in an appropriate position on the charging stations.
- A vehicle connector used for Direct Current (DC) charging shall be locked on a vehicle inlet if the voltage is higher than 60 V DC and the vehicle connector shall not be unlocked (if the locking mechanism is engaged) when hazardous voltage is detected through charging process including after the end of charging and in case of charging system malfunction, a means for safe disconnection shall be provided.
- If the output voltage exceeds the maximum voltage limit sent by the vehicle, the EVSE shall disconnect supply to prevent overvoltage at the battery.
- Whenever the vehicle connector is unlocked, the EVSE shall not energize the charging cable and the output voltage shall be less than 60 V.

The earth protection system for the EV charging stations is as follows [22]:

- All residual current devices (RCD) for the protection of supplies for the electric vehicle shall have a residual operating current of not greater than 30 mA; it has to interrupt all live conductors and neutral, performance at least equal to Type A, and be in conformity with IS 732-2018.
- All RCDs shall be permanently marked to identify their function and their locations in the charging station or socket outlet they protect.
- Each EVSE shall be supplied individually by a dedicated final sub-circuit protected by an overcurrent protective device complying with IEC 60947-2, IEC 60947-6-2, or the IEC 60269 series and the overcurrent protective device shall be part of a switchboard.
- Co-ordination of various protective devices shall be required.
- The discrimination (selectivity) shall be maintained between the RCDs (i.e. connection point and an upstream).
- All EV charging stations shall be supplied from a sub-circuit protected by a voltage-independent RCD and also provide personal protection that is compatible with a charging supply for an electric vehicle.

- All EV charging stations shall have an earth continuity monitoring system, and it has to disconnect the supply in the event that the earthing connection to the vehicle becomes ineffective.
- Earthing of all electric vehicle charging stations shall be as per IS 732.
- The cable may be fitted with an earth-connected metal shielding and the cable insulation shall be wear resistant and maintain flexibility over the full temperature range.
- A protective earth conductor shall be provided to establish an equipotential connection between the earth terminal of the supply and the conductive parts of the vehicle which shall be of sufficient rating to satisfy the requirements of IEC 60364-5-54.

Requirement of fire prevention for charging stations [22]:

- EV charging stations shall have firefighting systems as per the statutory regulations.
- Enclosure of charging stations shall be made of fire-retardant material with self-extinguishing property and free from Halogen.
- Fire detection, alarm, and control system shall be provided as per relevant Indian Standards.
- Power supply cables used in charging stations or charging points shall conform to IEC 62893-1 and its relevant parts.

10.7 BS 7671-2018 Requirements for Electrical Installations

Section 722 of this standard provides the requirements for electric vehicle charging installations. The requirements include the conductor arrangement and system earthing, protection against electric shock, selection and erection of equipment, etc. [23].

10.8 Conclusion

Regulatory bodies across the globe provide the regulations and standards for EV charging infrastructure requirements. The standards organizations like IEC, IEEE, ISO, SAE, ANSI, BIS, BSI, etc., provide the standards for various elements of the charging stations. These standards cover the EV charging station as a whole and include the design, construction, communication, and safety requirements. Generally, different standards cover the different elements of EV charging stations such as chargers, connectors, cables, switchgear, and safety. This chapter discusses the various IEC, ISO, IEEE, and SAE standards and their requirements for EV fast-charging stations.

References

1 IEC 61851-1 (2017). Electric vehicle conductive charging system – Part 1: General requirements.
2 Society of Automotive Engineers Standard J2894-1 (2011). Power Quality Requirements for Plug-In Electric Vehicle Chargers.
3 IEC 61851-21-1 (2017). Electric vehicle onboard charger EMC requirements for conductive connection to an AC/DC supply.
4 IEC 61851-21-2 (2018). EMC requirements for OFF board electric vehicle charging systems.
5 IEC 61851-23 (2014). DC electric vehicle charging station.
6 IEC 61851-24 (2014). Digital communication between a d.c. EV charging station and an electric vehicle for control of d.c. charging.
7 IEC 61980-1 (2020). Electric vehicle wireless power transfer (WPT) systems – Part 1: General requirements.
8 IEC 61980-2 (2023). Electric vehicle wireless power transfer (WPT) systems – Part 2: Specific requirements for communication between electric road vehicle (EV) and infrastructure.
9 IEC 62196-1 (2022). Plugs, socket-outlets, vehicle connectors and vehicle inlets – Conductive charging of electric vehicles – Part 1: General requirements.
10 IEC 62196-2 (2022). Plugs, socket-outlets, vehicle connectors and vehicle inlets – Conductive charging of electric vehicles – Part 2: Dimensional compatibility and interchangeability requirements for a.c. pin and contact-tube accessories.
11 IEC 62196-3 (2022). Plugs, socket-outlets, vehicle connectors and vehicle inlets – Conductive charging of electric vehicles – Part 3: Dimensional compatibility and interchangeability requirements for d.c. and a.c./d.c. pin and contact-tube vehicle couplers.
12 IEC TR 62933-2-200 (2021). Electrical energy storage (EES) systems – Part 2-200: Unit parameters and testing methods – Case study of electrical energy storage (EES) systems located in EV charging station with PV.
13 IEC 62893-1 (2017). Charging cables for electric vehicles for rated voltages up to and including 0,6/1 KV – Part 1: General requirements.
14 IEC 62893-4-1 (2020). Charging cables for electric vehicles of rated voltages up to and including 0,6/1 kV – Part 4-1: Cables for DC charging according to mode 4 of IEC 61851-1 – DC charging without use of a thermal management system.
15 IEC TS 62893-4-2 (2021). Charging cables for electric vehicles of rated voltages up to and including 0,6/1 kV – Part 4-2: Cables for DC charging according to mode 4 of IEC 61851-1 – Cables intended to be used with a thermal management system.
16 IEC 60364-7-722 (2018). Low-voltage electrical installations – Part 7-722: Requirements for special installations or locations – Supplies for electric vehicles.

17 IEEE Std 2030.1.1 (2021). IEEE Standard for Technical Specifications for a DC Quick and Bidirectional Charger for Use with Electric Vehicles.

18 IEEE Std 2836 (2021). IEEE Recommended Practice for Performance Testing of Electrical Energy Storage (ESS) System in Electric Charging Stations in Combination with Photovoltaic (PV).

19 SAE J1772 (2017). SAE Electric Vehicle and Plug-in Hybrid Electric Vehicle Conductive Charge Coupler.

20 ISO 17409 (2020). Electrically propelled road vehicles – Connection to an external electric power supply – Safety requirements.

21 CEA Technical Standards for Connectivity of the Distributed Generation Resources – 2019.

22 CEA Technical Standards for Measures relating to Safety and Electric Supply – 2019.

23 BS 7671 (2018). Requirements for Electrical Installations.

11

Fast-Charging Infrastructure for Electric Vehicles: Today's Situation and Future Needs

11.1 Batteries

11.1.1 Voltage

Presently, the nominal voltage range of battery packs in most of the light electric vehicles (EVs) is in the range between 300 and 420 V and heavy EVs is in the range of up to 800 V [1]. Based on the studies [2], a development or changes in the battery voltage trend is expected for light EVs, and the voltage of the DC bus is expected to be 800 V systems. An increase in the voltage has an impact on the conductor cross-section requirements. For carrying the same electric power, higher voltage level requires a lesser cable cross-section, and a significant weight reduction can be achieved. Also, the power loss in the cable (I^2R) is reduced. The battery voltage has a significant impact on the selection of the power converter topology for the charging station.

11.1.2 Improvements in Battery Chemistry

Table 11.1 gives the evolution of energy density, power density, charging cycle, lifetime, operating temperature range, efficiency, and recycling of Lead family, Nickel family, and Lithium family batteries by the year 2030.

From Table 11.1, the Lithium family is superior and dominates the Lead family and Nickel family in energy density, power density, charging cycles, and efficiency. The efficiency of the Lithium family is in the range of 90–95%. As compared with the Lead family and Nickel family, the efficiency of the Lithium family is 5% higher than the Lead family and 10% higher than the Nickel family. On the other hand, the charging cycle of the Lithium family is more than 10 000 cycles and which is more than double the time of the Lead family and Nickel family.

Fast-Charging Infrastructure for Electric and Hybrid Electric Vehicles: Methods for Large-Scale Penetration into Electric Distribution Networks, First Edition. Sivaraman Palanisamy, Sharmeela Chenniappan, and Sanjeevikumar Padmanaban.

Published 2023 by John Wiley & Sons, Inc.

Table 11.1 Evolution of battery technology by 2030 [3] / IEEE / Public Domain CC BY 4.0.

Features		Lead-based		Alkaline		Lithium-ion	
		Pb 2020	Pb 2030	Nix 2020	Nix 2030	Lithium ion 2020	Lithium-ion 2030
Electrochemical material	Cathode	PbO_2	PbO_2	B-NiOOH	B-NiOOH	NCM 111 (G.2a) NCM 523-622 (G.2b) LFP LMO LCO NCA	NCM 622-811 (Gen 3a) NCM811 HE-NCM HVS (Gen 3b) Solid State
	Anode	Pb, Pb + C	Pb, Pb + C	Cd, MH	Cd	LTO, C (G2a2b)	C + Si(5–10%) (Gen 3a) Si/C (Gen 3b)
Energy density (Wh/kg)	Cell system	24–48 23–45	30–60 35–55	28–50 24–43	30–55 38–50	60 – 250 20–140	300–450 80–400
Energy density (Wh/l)	Cell system	60–105 36–100	80–150 50–110	55–80 47–70	60–90 50–75	140–580 20 – 250	650–1100 100–1000
Power density (W/kg)	Cell system	34–448 41–400	80 – 505 65–450	80–225 68–180	100–240 80–210	210–1800 170–520	450–1100 250–700
Power density (W/l)	Cell system	91–880 76–840	120–920 72–900	112–400 95–350	120–460 100–380	470–2200 180–650	800–2500 600–1200
Lifetime	Cycles years	200–2500 10–25	1000–4800 10–25	3000 20	4000 20	>3500 10	>10.000 15–25
Op. temp. (°C)		−25 to +50	−25 to+ 50	−50 *to* +60	−50 *to* +60	0 to + 45 charge −20 to +60 disch. −30 to +55 LTO	−30 to +60
Efficiency (%)		67–85	>90	70–85	>85	>90	95
Recycling (%)		90	90	79	80–85	50	80–85

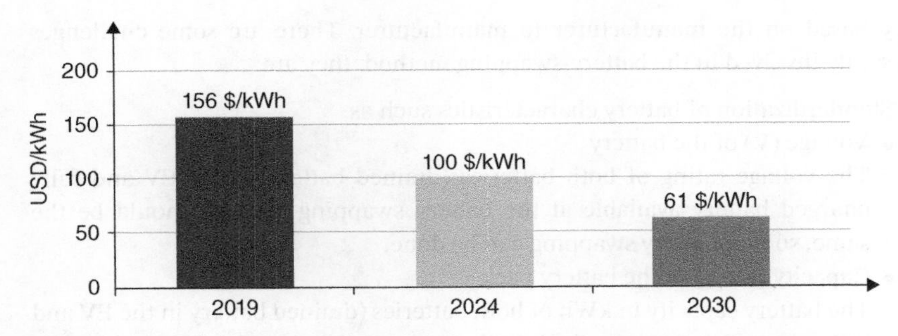

Figure 11.1 Li-ion battery pack average cost projection [3] / IEEE / Public Domain CC BY 4.0.

The lithium family also dominates in power density (W/kg) and it is in the range of 450–1100. The power density (W/kg) of the Lithium family is more than double the time of the Lead family and Nickel family.

Cost is one of the important factors for selecting the type of battery chemistry for a particular EV application. As compared with other chemistry, the cost of Li-ion batteries is slightly high. For large-scale adoption of Li-ion batteries in EV, the cost of Li-ion batteries has to be competitive with other battery chemistry.

Li-ion battery pack average cost projection for the year 2024 and 2030 is shown in Figure 11.1.

From Figure 11.1, as compared with 2019, the cost of a Li-ion battery is reduced to almost 35% for the year 2024 and reduced to almost 60% for the year 2030.

11.1.3 Standardization of Battery Ratings (Capacity, Voltage, and Dimensions) for Enabling Battery Swapping

Battery swapping is a method to replace the EVs' drained battery with a fully charged battery. Battery replacement can be done in a few minutes either manually or automatically by using robots like refilling the fuel in fuel stations for conventional vehicles. This method is saving the waiting time required for recharging the vehicle at charging stations. The battery swapping can be done at battery swapping stations and/or fast-charging stations with the facilities for battery swapping. Once the drained battery is replaced by a fully charged battery, the drained battery will be housed on the battery swapping station charging port. Once these drained batteries at battery swapping stations are fully charged, then the next EVs user can use it.

In order to attract users, EV manufacturers are providing different ranges (i.e. km/charge) for their vehicles. So, the battery capacity used in the vehicle will vary based on the manufacturer to manufacturer. There are some challenges presently involved in the battery-swapping method, they are:

1) Standardization of battery characteristics such as
   - Voltage (V) of the battery
     The voltage rating of both batteries (drained battery in the EV and fully charged battery available at the battery swapping station) should be the same, so that battery swapping can be done.
   - Capacity (kWh) of the battery pack
     The battery capacity in kWh of both batteries (drained battery in the EV and fully charged battery available at battery swapping station) should be the same, so that battery swapping can be done. In case of any difference in capacity, it would be difficult to swap the battery. If the EV battery rating is higher than the battery available at the battery swapping station, the EV user may not prefer to swap the battery.
   - C-rating
     C-rating of the battery is used to describe the charging and discharging characteristics of the battery. C-rating of both batteries is to be the same.
   - No of cells or modules in a battery pack
     A battery pack used in EV consists of multiple cells or modules at achieving the rated voltage and capacity. The number of cells or modules of both batteries is to be the same.
2) Rating of battery capacity is varied from manufacturer to manufacturer
   - In order to attract users, every EV manufacturer is offering a different travel range or mileage. So, the battery capacity and its characteristics are varied from manufacturer to manufacturer. Hence, the requirements of battery swapping stations are increased to cater to all EV manufacturer's vehicles.
3) Battery technology is varied from manufacturer to manufacturer
   - The battery technology (Li-ion, lead acid, NMH, etc.) used in the EVs also varied from manufacturer to manufacturer.
4) Battery pack dimensions varied from manufacturer to manufacturer
   - For the same battery pack capacity, different manufacturers are providing the battery pack at different dimensions. If the particular manufacturer's battery pack is not available at the battery swapping station, then the EV users are not able to swap the battery due to battery pack dimension issues.
5) Cost of the battery is varied from manufacturer to manufacturer
   - The cost of the battery is varied from manufacturer to manufacturer for the same battery technology and same rating.

A country-specific battery-swapping policy is essential for enabling the battery-swapping option to the maximum extent. The policy must address the present challenges like standardization of battery capacity, voltage ratings, dimensions, cost, etc.

11.2 Distributed Energy Storage System and Grid-Friendly Charging

The batteries in the EVs act as an energy storage system (ESS) and it provides the required power and energy to run the vehicles. Whenever these EVs are in standstill mode or not in operation, the energy stored in the EV batteries can be fed back to the grid. In simple words, EVs act as a distributed ESS. Most of the private noncommercial vehicles have more nonworking conditions and these vehicles can feed the energy back to the grid. Feeding the energy back to the grid is called Vehicle to Grid (V2G) integration. The main advantages of using V2G methods are EVs are charged at off-peak hours at a lesser energy cost and discharge the energy (feed the energy) back to the grid during the peak hours at higher energy cost. Some of the other advantages of V2G are:

- Meeting the required peak demand without much capital investment
- Power quality enhancement
- Voltage support
- Power balancing and frequency control

If a large number of vehicles are charging during peak hours, it will have an impact on the distribution system power demand as well as physical distribution infrastructure or equipment. Time of Use (ToU)-based tariff structure of DISCOMs plays a major role in shifting the EV charging time to off-peak hours and V2G integration. ToU tariff provides different tariff structures for different grid operating conditions. If the grid is operating with lesser loads during the off-peak hours, the energy cost is low and if the grid is operating with higher loads during the peak hours, the energy cost is high. The EV users can take this ToU tariff as an advantage and they can charge their vehicle during an off-peak hour at a lesser energy cost and sell back to the grid at a higher cost.

Role of DISCOM: providing the attracting ToU tariff structure

Role of EV users: follow the ToU tariff structure and charge the vehicle during off-peak hours

Role of researchers: providing better technology for communication and control between the vehicle and distribution grid operator. Because large numbers of EVs will be connected across the distribution system for V2G integration during peak hours. Communication and control between the vehicle and distribution grid operator are essential for smoother grid operation.

11.3 Ultrafast Chargers

Generally, fast chargers available in the market are up to 400 kW power rating. Heavy vehicles used for commercial purposes have a large battery capacity and charging these heavy vehicles in lesser duration demands a high-power charger whose power rating is more than 400 kW. These chargers are also called ultrafast chargers. Ultrafast chargers will reduce the waiting period required at the charging station to charge the EVs. IEEE is an international professional body that is developing a standard for ultrafast charging.

The IEEE P2030.1.2 – Standard Technical Specifications for Ultra-High-Power Electric Vehicle Chargers is presently in the development stage. This standard defines the technical requirements for the design of safe, compact, and reliable ultra-high-power DC chargers for EV charging. The power rating of ultrafast chargers considered in this standard is 500 kW or above. Also, this standard covers the dynamic current control, and bidirectional charging for Vehicle to Building (V2B), V2G, and Vehicle to Home (V2H) applications. This standard is applicable to both light-duty and heavy-duty vehicles [4].

11.4 Interoperable Features

Presently, multiple plugs, connectors, or couplers are available in the market like CCS, GB/T, SAE, CHAdeMO, etc. The fast charger with a CCS connector is able to charge the EV which is suitable for a CCS connection. This fast charger is not suitable for a CHAdeMO connection. Similarly, a fast charger with a CHAdeMO connector is able to charge the EV which is suitable for a CHAdeMO connection; this fast charger is not suitable for a CCS connection or GB/T connection. Hence, interoperability is essential for enabling the different types of connectors suitable for EVs.

11.5 Charging the Vehicle While Driving (Wireless Charging)

Wireless charging is an emerging topic in the EV ecosystem. This provides contactless energy transfer from the source to the EV batteries. At present, wireless charging to an EV while driving on road is not a mature technology. Many researchers are working on this wireless charging for EVs while driving on the road.

The advantages of using the wireless charging method while driving the EVs on roads are [5, 6]:

- Enabling long-distance travel without any waiting period for charging.
- Vehicle range anxiety is not a major concern.

- Avoiding the queue at charging stations.
- Can be able to start the journey even though the battery SoC is in low condition.

There are some technical challenges involved in wireless charging while driving:

- Infrastructure development: wireless charging infrastructure to be developed throughout the road in the planning area.
- It is difficult to establish the infrastructure on the already existing road.
- Providing the electricity throughout the planning area.
- Accounting for the energy consumed by the vehicle.

11.6 Conclusion

EVs are an absolute necessity due to the rapid degradation of fossil fuels and are free from environmental pollution. The fast-charging method is used to charge the EVs with higher power ratings and usually it will take 15–30 minutes to recharge the battery up to 80% SoC. This method significantly reduces the waiting time of EV users at charging stations. So, the fast-charging method is a widely adopted method to recharge EVs used for commercial purposes including freight transfers. The typical power rating of a fast charger nowadays is available in the range of 50–150 kW rating. The C-rating of the EV battery plays an important role in charging. A lesser C-rating of the battery will take more time for charging even though if it is charged through fast chargers. This chapter discussed the battery evolution for the year 2030, ultrafast charger, EV as the distributed ESS, ToU for grid-friendly charging, etc.

References

1 "Voltage classes for electric mobility," ZVEI, Die Elektroindustrie, Frankfurt, Germany, Tech. Rep., 2013. https://www.zvei.org/en/voltage-classes-for-electric-mobility (accessed 29 October 2022).

2 Jung, C. (2017). Power up with 800-V systems: the benefits of upgrading voltage power for battery-electric passenger vehicles. *IEEE Electrification Magazine* 5 (1): 53–58.

3 Aretxabaleta, I., De Alegria, I.M., Andreu, J. et al. (2021). High-voltage stations for electric vehicle fast-charging: trends, standards, charging modes and comparison of unity power-factor rectifiers. *IEEE Access* 9: 102177–102194.

4 IEEE P2030.1.2 – Standard Technical Specifications for Ultra-High-Power Electric Vehicle Chargers. https://standards.ieee.org/ieee/2030.1.2/10485 (accessed 15 March 2023).
5 Chen, H., Zhou, S., Hui, Y., and Hui, H. (2018). Dynamic charging optimization for mobile charging stations in internet of things. *IEEE Access* 6: 53509–53518.
6 Sivaraman, P., Sharmeela, C., and Logeshkumar, S. (2021). Charging infrastructure layout and planning for plug-in electric vehicles. In: *Cable Based and Wireless Charging Systems for Electric Vehicles* (ed. P. Sanjeevikumar), 1–24. IET.

12

A Review of the Improved Structure of an Electric Vehicle Battery Fast Charger

Mohammad Zand, Mostafa Azimi Nasab, Samaneh Rastgoo, and Morteza Azimi Nasab

Department of Electrical Engineering, IT and Cybernetic, University of South-Eastern Norway, Kjølnes Ring 56, 3918 Porsgrunn, Norway

12.1 Introduction

Fossil fuel depletion, environmental concerns about energy sources for instant earth warming, and deteriorating weather are all reasons to replace fossil fuels [1]. Transportation is one of the most important parts of fossil energy consumption, so recent research has focused on developing efficient, clear, and safe transportation systems. Today's electrical transportation is characterized by increased operation and sustainability in the transportation field and decreased environmental pollution [2]. Shortly, electrical vehicles (EVs), hybrid electrical vehicles (HEVs), and fuel cell-based vehicles are being offered to replace conventional vehicles, and we can coordinate that with renewable energy to manage energy in the grid [3]. Industries and universities have recently discovered that energy-efficient EVs and HEVs reduce carbon dioxide (CO_2) emissions, according to renewable energy development and incentive methods to increase efficiency.

However, all reports show that the main reasons for preferring electric vehicles to internal combustion vehicles are time spent, charging time, and traveled distance [4]. Due to the future share market expansion of electric vehicles, it is necessary to decrease the charging time near the car refueling at a fuel station and increase the travel distance during each charge. In this chapter, we investigate the structure of improvement for battery fast charging in four vehicles.

Fast-Charging Infrastructure for Electric and Hybrid Electric Vehicles: Methods for Large-Scale Penetration into Electric Distribution Networks, First Edition. Sivaraman Palanisamy, Sharmeela Chenniappan, and Sanjeevikumar Padmanaban.

Published 2023 by John Wiley & Sons, Inc.

12.2 Types of Battery Charging

There are two types of vehicle charging: on-board and off-board. The on-board charging needs AC single-phase electricity, and there is no need to access a charging station. The on-board charging output power is low, and the different voltage levels are shown in the Table 12.1.

The battery charging duration for on-board charging is longer than off-board charging, and because of the low level of demand, there are fewer elements in the on-board charging structures.

Because of the reduced use of bottoms and the low cost of the passive and active elements, these chargers are small and light. The off-board chargers are bigger and need three-phase electricity for operation. According to the high output demand of these chargers, the elements and bottoms are powerful as listed in Table 12.2.

Table 12.1 Different AC voltage levels of the on-board charging.

Level 1	120 V, 1.4 KW 120 V, 1.9 KW PHEV: 7 h (SoC 0% to full) BEV: 17 h (SoC 20% to full)
Level 2	240 V, 19.2 KW (80 A) 3.3 KW Charging time PEV: 3 h (SoC 0–100%) BEV: 7 (SoC 20% to full) 7 KW Charging time PEV: 1.5 h (SOC 0% to full) BEV: 3.5 h (SOC 20% to full) 20KW Charging time PEV: 22 min (SOC 0% to full) BEV: 1.2 h (SOC 20% to full)
Level 3	>20 kW, single-phase, and three phase

Table 12.2 Different DC voltage levels of the off-board charging.

Level 1	UP TO 36 KW (80 A) @ 200–450 V DCPHEV: 22 min (SOC 0% to full)BEV: 1.2 h (SOC 20% to full)
Level 2	UP TO 90 KW (200 A) @ 200–450 V DCPHEV: 10 min (SOC 0% to full) BEV: 20 min (SOC 20% to full)
Level 3	UP TO 240 KW (400 A) @ 200–600 V DCBEV (only): 10 min (SOC 0% to 80%)

Distinct types and powers are produced and supplied by electric vehicle (EV) chargers. The EV battery chargers must have almost one power factor (PF), high efficiency, low cost, a simple controlling method, and high density.

12.2.1 Li-Ion Battery Charger Algorithm

Li-ion batteries charge through ion transfer between two electrolytes. Common charging methods include:

- Constant voltage–current charging method
- Constant current multilevel charging method
- Incremental charging method
- Pulse charging method

The best charging method reaches the maximum SoC with high efficiency and charges the battery faster.

12.2.2 Constant Voltage–Current Charging Method

As shown in Figure 12.1, Li-ion batteries include diameters charging, constant current and voltage, and charging cut. This method is the best charging method.

In the first level of charging batteries with constant current, the voltage drop of the battery cells is fixed at the constant current level [5]. Until reaching the diameter threshold voltage, the battery cell voltage is increased. In the second level, batteries charge at a constant current. This charging method decreases the charging duration. The electrode reaction raises the cell's voltage when a battery is charged with a constant current. When the voltages go up, the cell double resistance will decrease and shorten the duration; therefore, the time at this level

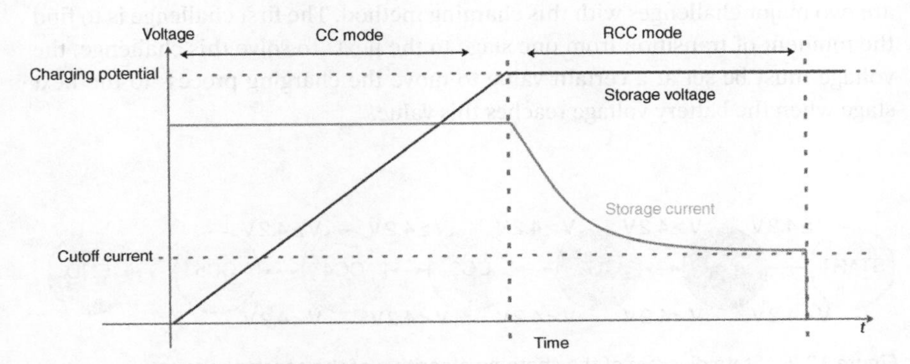

Figure 12.1 Constant voltage–current charging method level.

decreases. With increasing time at a constant voltage level, the duration of full charges will increase. After this constant current level, a constant voltage level avoids charging current. In the final level, the battery is typically charged in two ways: one with the decreasing current into the minimum amount and another with decreasing time into the minimum amount, or a combination of both. Charging will be stopped with a decreasing current after reaching the charge current's minimum amount. The other extra systems in chargers for protection are considered. For instance, the battery degree is usually 0°–45°; if it is out of the bound, the charging will stop. Going out of this range will cause the battery to heat during charging, and the pressure and temperature will increase, causing the battery to swell. Due to the direct relationship between the pressure and temperature of the battery, with an increase in temperature, the pressure of the battery also increases, and leaving the temperature range reduces the life and efficiency of the battery. The thermistor is used to measure the temperature of the battery in lithium batteries. The resistance is measured between the negative end of the battery and the thermistor terminal. The charging operation is stopped if the resistance is higher than the allowed value. Due to the equalization of the voltage in each cell and the demand for proportional charging of the cells, monitoring the voltage of each cell is one requirement of these batteries while connecting the series of lithium batteries.

12.2.3 Constant Current Multilevel Charging Method

The state diagram of the constant current multistage charging algorithm is shown in Figure 12.2 [6]. Eliminating the constant voltage area increases battery life and reduces charging time. This charging method consists of five stages: the constant current stage is divided into five stages, and the constant voltage stage is removed. The battery will be charged in each stage with a different constant current. There are two major challenges with this charging method. The first challenge is to find the moment of transition from one stage to the next. To solve this challenge, the voltage must be set at a certain value to move the charging process to the next stage when the battery voltage reaches this value.

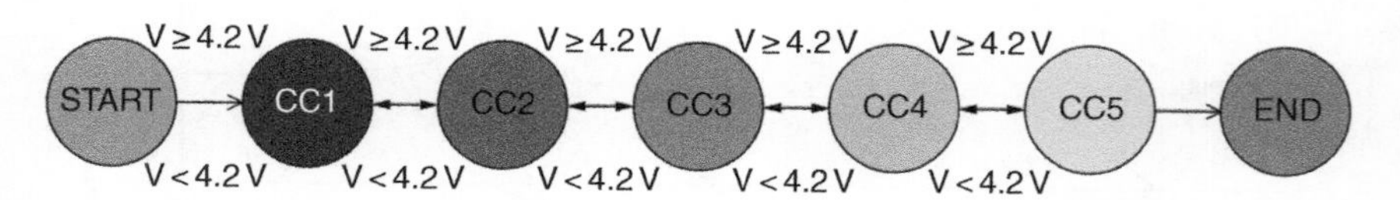

Figure 12.2 State diagram of the charging algorithm of the constant current multistage method.

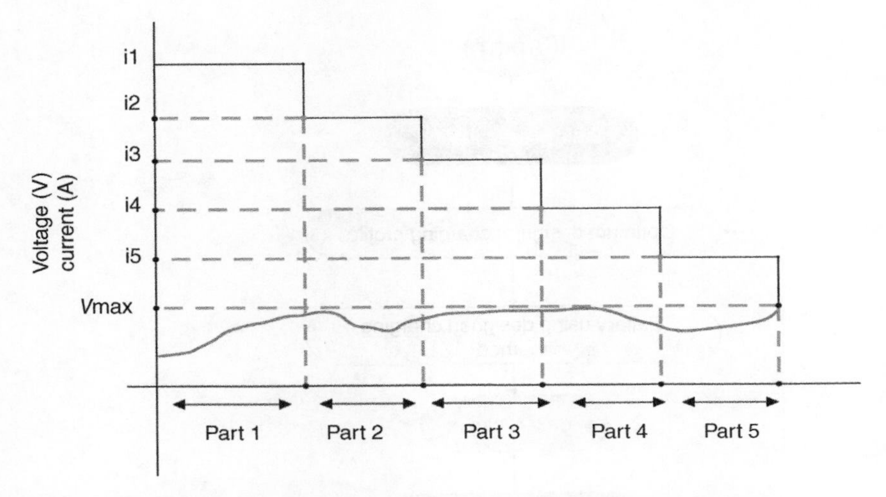

Figure 12.3 Steps of battery charging by the constant current–voltage method.

The second challenge is determining the battery's current in each charging stage. In this method, the charging time is reduced. The flowchart and stages of constant multistage current charging are presented in Figures 12.3 and 12.4.

12.2.4 Method of Incremental Charging

In this method, the battery must be completely discharged to perform the charging process. In the constant current mode, the maximum charging voltage should be slightly higher than the battery voltage (for example, 4.3 V, which is 0.1 V more than 4.2 V). The battery discharge process should be completed in an abbreviated time. After this stage, the charging algorithm is conducted as the standard constant current–voltage charging algorithm. In this method, the battery must be completely discharged before charging. Therefore, it needs a discharge circuit, which increases the cost and number of parts. According to the tests, the initial voltage does not negatively affect the battery life. This high initial voltage method can charge the battery faster than the constant current–voltage method.

12.2.5 Pulse Charging Method

The flow chart of the pulse charging method is shown in Figure 12.5. In this algorithm, electrochemical characteristics are used to increase battery life. In time periods of 1 and 2 t, a lithium battery will be charged with two different current levels (one being five times the other). The battery's open-circuit voltage is

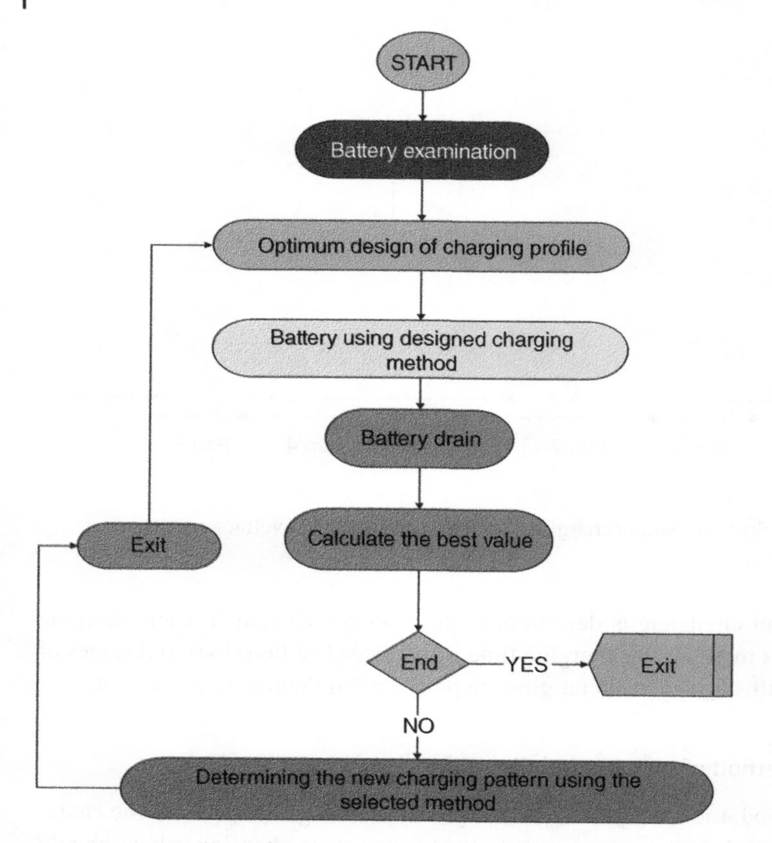

Figure 12.4 Charging flow chart by the multistep constant current method.

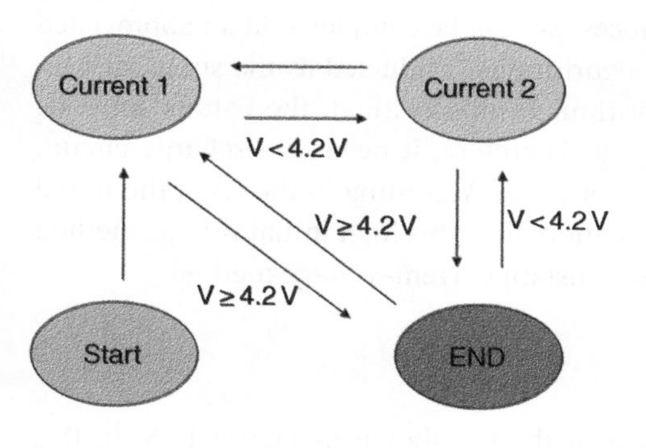

Figure 12.5 State diagram of the pulse method charging algorithm.

compared with the reference voltage during this charging process, and when it reaches a certain value (for example, 4.2 V), this charging process is completed. In the pulse charging method, the current can flow in one direction and is usually unidirectional. For it to be bidirectional, the cost of the charger must be increased,

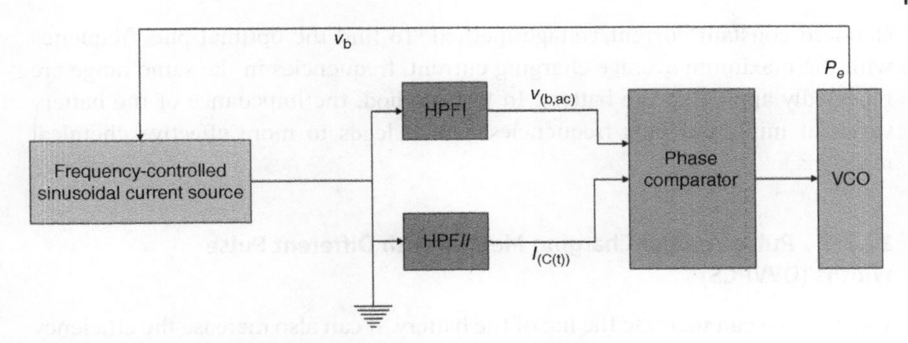

Figure 12.6 Block diagram of the sinusoidal pulse charging method.

which is the disadvantage of this method. The algorithms of this charging method can be divided into two groups, which will be examined further.

12.2.6 Sinusoidal Pulse Charging Algorithm

The block diagram of this method can be seen in Figure 12.6. To find the optimal frequency, it is necessary to function as follows: first, a sinusoidal voltage with an offset should be produced. This generated voltage is given to the battery, and the current flowing from the battery is calculated.

If these two voltage waves and the measured current have a phase difference, it means that the battery is not optimally charged and it must reach a resonant frequency where the battery is like a resistor, the voltage signal is sent again with a different frequency, and the current waveform is received, and then the phase difference is measured according to the previous procedure. This cycle will continue until the optimal frequency is found. Therefore, we reach the resonant frequency when the phase difference between the current passing through the battery and the voltage is zero. During charging, you can change the optimal frequency and impedance obtained from the resonance frequency calculation method.

12.2.7 Using a Different Frequency Pulse Charging Method (VFPCS)

According to the electrochemical characteristics of the battery, the charging time can be reduced. The variable frequency pulse charge system (VFPCS) method is used to improve the performance of conventional Plus charging. This experimental method was obtained by trial and error [7, 8]. The variable-frequency pulse charging method can obtain the optimal charging frequency and increase the battery charging speed. The Plus frequency is varied within a certain range in this method to obtain the optimal frequency with the lowest battery resistance. According to studies, the charging speed in this method is 21% faster than the standard constant current/voltage method. To find the optimal plus frequency with the maximum average charging current, frequencies in the same range are repeatedly applied to the battery. In this method, the impedance of the battery varies at minimum plus frequencies, which leads to more effective chemical reactions.

12.2.8 Pulse Voltage Charging Method with Different Pulse Widths (DVVPCS)

This method can increase the life of the battery. It can also increase the efficiency and speed of charging and be a suitable charger for the battery. The pulse voltage charging method with variable pulse width increases the efficiency by 3.4% and the charging speed by 14% compared to the constant current–voltage method [9]. In Table 12.3, the advantages and disadvantages of each battery charging method (constant current–voltage charging method, constant current multistage charging method, incremental charging method, and pulse charging method) are summarized and presented.

12.2.9 An Overview of Lithium-Ion Batteries

Rechargeable lithium-ion batteries are the most suitable for EV applications due to their maximum energy density and longer life cycle. The formula for calculating the battery volume is given in Eq. (12.1).

$$\text{Volume of battery} = \frac{(\text{Device longevity})(\text{Average power})}{\text{Energy density}} \tag{12.1}$$

Table 12.3 Comparison between the proposed algorithms.

Advantage	Disadvantage	Charging algorithm
Simple control	Increasing charge time and constant current–voltage battery temperature	Complex control
Increasing the efficiency and battery life	Complex control	Multistage constant current
Fast charging up to 60%	Increases the cost and requires discharge circuit	Booster
Increasing battery life and fast charging	Not fully charging the battery and being unidirectional	Pulse

A lithium-ion battery is a rechargeable battery made up of positive and negative electrodes separated by a separator. The electrolytes used in lithium-ion batteries are $6LiBF_4$, LiPF, 4LiClO, etc. [10]. It provides a conductive medium for electron flow. Lithium ions move from the positive electrode through the electrolyte to the negative electrode during charging. Therefore, in the external circuit, the electrons flow in the same direction. During discharge, the opposite happens. Lithium-ion batteries are lightweight compared to lead-acid batteries. Therefore, lithium-ion batteries are portable batteries and are used for EV applications [11]. The charging capacity increases with the increase in voltage, and the current is fixed in the constant current operation mode. When the voltage reaches its saturation value, the current decreases exponentially. During discharge, constant voltage and constant current are maintained, and after a while, with a slight increase in current in the load, the voltage level decreases slightly. It can supply the load's energy up to the specified cutoff voltage. Lithium-ion batteries have good efficiency over a wide temperature range, an exceptionally low self-discharge capacity, and a long life [12].

12.2.10 Performance Comparison with Other Batteries

The performance characteristics of lithium-ion batteries compared to other types of batteries are presented in Table 12.4. Lithium-ion batteries have a good energy density (110–160 Wh/kg) and have optimal power and energy density for EV applications. In addition, the operating temperature range for lithium-ion batteries is high, which is an additional advantage for EV applications. The weight of batteries is the main challenge for electric vehicles.

Therefore, lithium-ion batteries are lighter and smaller than other batteries. The discharge capacity of lithium-ion batteries is linear compared to other batteries [13].

Table 12.4 Performance analysis of lithium-ion batteries with other batteries.

Battery kind	Li-ion	Ni-Cd	Zn-Br	Ni-MH	Lead-acid
The energy density (Wh/Kg)	110–160	35–54	40–85	70–100	30–45
Power density	1800	150	—	250–1000	180
Temperature (Celsius)	−20 to 60	−40 to 60	−20 to 60	−20 to 60	−20 to 60
Working voltage (V)	3.6	1.25	1.67	1.25	2
Self-discharge	Low	Medium	Low	Much	Low
Energy efficiency	80	80	80	75	70

12.2.11 Lithium-Ion Battery Control System (BMS)

Batteries are part of the main component of electric vehicles that represent sustainable transportation. Lithium-ion technology is the most desirable technology for batteries in electric vehicles. However, the research field for this technology is still exceptionally large, including the proper selection of materials or cells for energy storage, implementation of electronic circuits, and the efficient operation of batteries using appropriate algorithms. The EV system must be safe to improve the battery's efficiency and its safe operation. For this purpose, operations such as state monitoring, data monitoring and analysis, temperature control in the battery, and cell voltage balance must be performed. Battery-management system (BMS) protects the battery from damage, increases the health of the battery, and also monitors that the cells are ready for use at all times. To protect the proper operation of the vehicle, the BMS must communicate with the vehicle controller and the engine controller. Two communication ports are usually used, one through the data bus and the other through the local control network.

12.2.12 Cell Control

The battery is formed by connecting a series of cells. Therefore, having a battery cell monitoring system in the BMS is important. To increase battery health and efficiency, it is necessary to monitor individual cells. In addition, the cell monitoring system (BMS) analyzes and controls battery parameters such as temperature dynamics in each battery cell, current, and voltage. Each battery cell has its temperature and current sensor. Sensor signals communicate with the controller module, and then the signals are transferred to the central controller module. These signals are used for control and protection. The data calculated by the battery cell control system is used to calculate the state of charge (SoC). It also controls the unbalanced voltage that occurs during charge and discharge cycles. A good battery cell monitoring system monitors battery parameters such as SoC, discharge function (SoF), and capacity remaining (SoH). Therefore, battery cell monitoring improves battery performance, reduces errors, and protects the battery [14].

12.2.13 Checking Input and Output Current and Voltage

A simple voltage division network and BMS controller perform the control process. While the vehicle is off, it measures the actual output voltage of the battery in the same state, and if the vehicle is on, it measures the battery charging voltage. Changing the voltage and current from the battery may damage the battery. Therefore, it is necessary to control the basic electrical parameters of the battery to protect it against overvoltage strategies [15]. It also stores voltage and current values and uses them for other applications.

12.2.14 Battery Charge and Discharge Control

The performance of batteries depends on their charging and discharging. For proper or optimal performance, each battery has its operating voltage range. In the ideal state for lithium-ion batteries, it is about 2.5–4.2 V. The charge and discharge cycle may be different in some physical conditions. The efficiency of the battery system can be calculated from Eq. (12.2).

$$\text{Efficiency} = \text{Discharge energy} - \text{Charging energy} \tag{12.2}$$

12.2.15 State Estimation

The estimation of battery states includes the SoC, the state of health (SoH), and the state of operation (SoF).

12.2.16 State of Charge

The SoC means the fraction of the charge in the battery until full charge and is expressed as a percentage. Generally, battery SoC can be calculated by the ratio of real-time charges to the full battery capacity charges. SoC estimation can find EV mileage and the BMS control index. Equation (12.3) is used to calculate a battery's SoC.

$$\text{SoC} = \frac{Q_t}{Q_n} \tag{12.3}$$

where,

Q_t is real-time charges present in the battery
Q_n is full battery capacity charges

SoC is determined by algorithms such as the battery model-based method, the open circuit voltage method, the sliding mode observation method, the discharge test method, the normal Kalman filter method, the neural network model method, the fuzzy logic method, and the integral method. Ampere hours, thermal calibration method, weighted fusion algorithm, developed Kalman filter method, etc., are estimated. The main application of SoC estimation is to find the driving distance in electric vehicles, and it also helps determine the time to use the fossil fuel engine in hybrid vehicles. SoC estimation can be done both in a steady and dynamic state. The comparison of different SoC estimation methods is given in Table 12.5.

Table 12.5 Comparison of state of charge estimation methods.

Method	Limitation	Main advantage	Input
Discharge test	It takes much time, and there is a waste of energy	Accurate, easy to implement	Remaining charging capacity
Integral ampere-hours	Batteries under variable or dynamic conditions are not accepted	Easy to implement, accurate	Self-discharge rate
Open circuit voltage	Adaptable in situations where SOC is maximum or minimum	Accurate flow measurement is required	Voltage at rest
SOC based on battery model	Not suitable for steady-state batteries	Independent of SOC value	Current, voltage
Neural network	Requires training data	Suitable for all batteries	Cumulative charge, the initial amount
Fuzzy logic	It is not precise	Good efficiency of	Current, voltage
Weighed algorithm fusion	Instability when the weighting coefficient is not suitable	It is not sensitive to initial values	Self-discharge rate, SoC initial value
The Kalman filter	Requires computational data	Accurate, dynamic, not sensitive to noise	Rest time, the initial value of SoC
View in sliding mode	It is nonlinear and difficult to implement	It is not noisy. Not sensitive to noise	Self-discharge rate

12.2.17 State of Health (SoH)

This index compares the current state of the battery with the absolute state of the battery, which depends entirely on the resistance of the internal cell. The SoH of the battery depends on the usable capacity of the battery and the total capacity of the battery. SoH is used to find the health of lithium-ion batteries. The SoH of a battery is calculated from the equation.

$$\mathrm{SoH}(\%) = \frac{Q_{\mathrm{max}}}{Q_{\mathrm{n}}} \times 100 \tag{12.4}$$

where,

Q_{max} is maximum charge available in the battery
Q_n is the rated capacity of the battery

1SoC and 2SoC are the states of charge at each start and end of charging. If the battery's SoH index reaches 80% of its rated value, it dries up or dies. SoH indicates

Table 12.6 Comparison of battery failure methods for SoH estimation.

Method	Function, no data	Real-time performance	Precision	Conformity
Models of electrochemical techniques	Better	Better	Excellent	Excellent
Equivalent circuit models	Good	Good	Better	Very weak
Semiempirical-based models	weak	Better	Excellent	Very weak
Analytical models	Not satisfactory	Not satisfactory	Satisfying	Not satisfactory
Statistical models	Very weak	Better	Better	Better

the battery's failure time. There are diverse ways to find the battery failure rate. The characteristics of different methods are mentioned in Table 12.6.

12.2.18 Mode of Operation (SoF)

The SoC, health, temperature, and charge and discharge data determine the SoF index. For a battery used in an energy storage system, the operating mode depends on the energy available in the battery and the highest possible amount of energy that can be stored in it.

In the EV system, the battery must meet the engine's needs, so the state of SoF must be considered. The quality index of the SoF for the discharge curve in terms of SoC and SoH during operation is used with adaptive control technology to identify and estimate the state of the batteries. Moreover, signals are sent to the controllers of electric vehicles to protect them and increase their efficiency.

12.2.19 Battery Protection

All charge storage systems use batteries for the same purpose. This function is also performed in electric cars using battery cells connected in series. The bad condition of the batteries is due to the increase in the temperature of the cells, the problems with the cells, and their construction. These conditions may lead to a decrease in the efficiency and performance of batteries and battery life. Chemical reactions in the battery, especially in each cell, produce or release heat and can lead to a fire accident involving the entire battery system. Hence, charge controllers are especially important for protecting battery cells in an array and the safe operation of energy storage devices. For this reason, to reduce this problem, most batteries use common organic solvents as battery electrolytes.

Cell balancing a battery is a combination of individual cells that are connected in series. Since several cells are used in the battery, two cells may not be the same in terms of discharge rate, SoC, heat, etc., and may have slight changes. Cell imbalance has many effects on the BMS. Cell-balancing circuits are usually divided into active and passive categories. The passive method is also known as resistor bleeding balancing, and switches balance the cell. Conventional algorithms used in cell balancing are the voltage control method and the SoC algorithm. In the voltage balancing algorithm, the unbalanced cell is identified first. A cell whose voltage difference from the average voltage of the cells is greater than the threshold voltage is called an unbalanced cell. The unbalanced cell is corrected and then used or replaced with a new cell. In a state-of-charge-based algorithm, state-of-charge values smaller than SoC control the threshold. Therefore, balancing and equalizing the cells protects the battery from damage, explosion, and improper operation. Therefore, it is important to adopt cell balance in the battery management system. The characteristics of battery charge equalization techniques and their advantages and limitations are discussed in Table 12.7. The coupled inductor is effective in charge equalization and has a fast equalization speed. Although the performance of boost and boost-buck converter charging techniques is good, their control strategies are complex and have switching losses.

Table 12.7 Characteristics of battery charge equalization techniques.

Charge equalization technique	Disadvantages	Advantages	Efficiency
Coupled inductor	High-current switching is required, as are capacitor filters	Two-way, high draw speed	Top
Boost converter	It requires an intelligent control strategy	Double-sided, extremely high draw speed, excellent efficiency	Top
Increase–decrease converter	Expensive; strong keys are required	Simple processing, low loss, high efficiency	Top
Two-way flyback converter	Implementation complexity	Two-way, low losses, minimal stress	Medium
Complete shunt	Wide voltage range for converters	Low power loss, low stress, equal speed	Medium

12.3 Temperature and Heat Control

Heat management matters in battery management systems. The production of heat leads to a decrease in efficiency and a decrease in the life cycle of the battery. The reasons for heat production are associated with the performance of lithium-ion batteries. The heat generated in the battery is due to reaction, resistance, reversible heat, and heat from the external connection terminal. The heat must be dissipated quickly, or the battery temperature will rise. The heat caused by the reaction is created due to the exchange of electrons from one electrode to another. The total bidirectional heat generated is related to the entropy changes at the two electrodes. The external terminal connection resistance produces heat due to the resistance between the cell terminals and the surface connection. Thermal control is important for increasing phasing, cycle life, and battery charge capacity. Therefore, the heat can be managed using the cooling system in the battery. Some methods used in heat management are active cooling systems, liquid cooling, air cooling, direct liquid cooling, and indirect contact cooling. Batteries work best at room temperature, between 25 and 40°C. Operating the battery at a certain elevated temperature improves its performance, but overall, it reduces the life of the battery. The performance of batteries decreases with a decrease in temperature (0–40°C). The temperature of the battery can eventually reach more than 1000°F. However, when it reacts with the oxygen in the atmosphere at that temperature, the electrolyte can ignite or even explode. Therefore, a heat management system must be used in EVs. Along with battery discharge and heat management, the SoC balancer also plays a key role in effectiveness, efficiency, and safety.

12.3.1 Examining the Charger Structure

Table 12.8 shows some examples of advanced DC fast chargers available on the market. The most advanced DC chargers convert the three-phase AC voltage (480 V) into the desired DC voltage using power electronic converters in two stages: the first stage with an AC–DC converter with power factor correction (PFC) and AC input voltage. Three phases produce medium DC voltage, and the second stage, with a DC–DC converter, converts the medium DC voltage to the adjustable DC voltage required to charge EV batteries. Galvanic isolation between the grid and the EV battery can be done in two ways: the first method is to use a line-frequency transformer before the AC–DC converter, according to Figure 12.7a. The second solution for isolation is to use a high-frequency transformer inside the DC–DC converter, according to Figure 12.7b. If a single module charger does not meet the power requirements of the DC fast charger system, several identical modules are used in parallel to increase output power, as shown in Figure 12.7c,d.

Table 12.8 Some examples of DC fast chargers available on the market.

Model and manufacturer	ABB TERRA HP	EVTEC ESPRESSO and CHARGE	TESLA SUPERCHARGER	PHIHONG INTEGRATED-TYPE	TRITIUM VEEFIL-RT	ABB TERRA 53
Power	350 KW	150 KW	1350 KW	120 KW	50 KW	50 KW
Supported protocols	SAE Combo-1 CHAdeMO O 1.2	SAE Combo-1 CHAdeM O 1.0	Super-charger	GB/T	CCS Type 1 and 2 CHAdeM O 1.0	CCS Type 1 CHAde MO 1.0
Input voltage	$4V_{ac} \pm 10\%$ 00	$4V_{ac} \pm 10\%$ 00	$3^\wedge 0 - V_{ac}\ 4^\wedge 0$	$3V_{ac} \pm 15\%^\wedge 0$ $4^\wedge 0V_{ac} \pm 15\%$	$3^\wedge 0 – V_{ac}\ 4^\wedge 0$ $600–900V_{dc}$	$4^\wedge V_{ac}\ 0$
Output voltage	150–920 V	170–500 V	50–410 V	200–750 V	200–500 V 50–500 V	200–500 V 50–500 V
Output current	375 A	300 A	330 A	240 A	125 A	120 A
Maximum efficiency	95%	93%	91%	93.5%	92%<	94%
Capacity	18941	15^11	10471	5911	4951	75^1
Weight	1340 kg	400 kg	600 kg	240 kg	165 kg	400 kg
Time to travel 200 miles	10 min	24 min	27 min	30 min	72 min	72 min

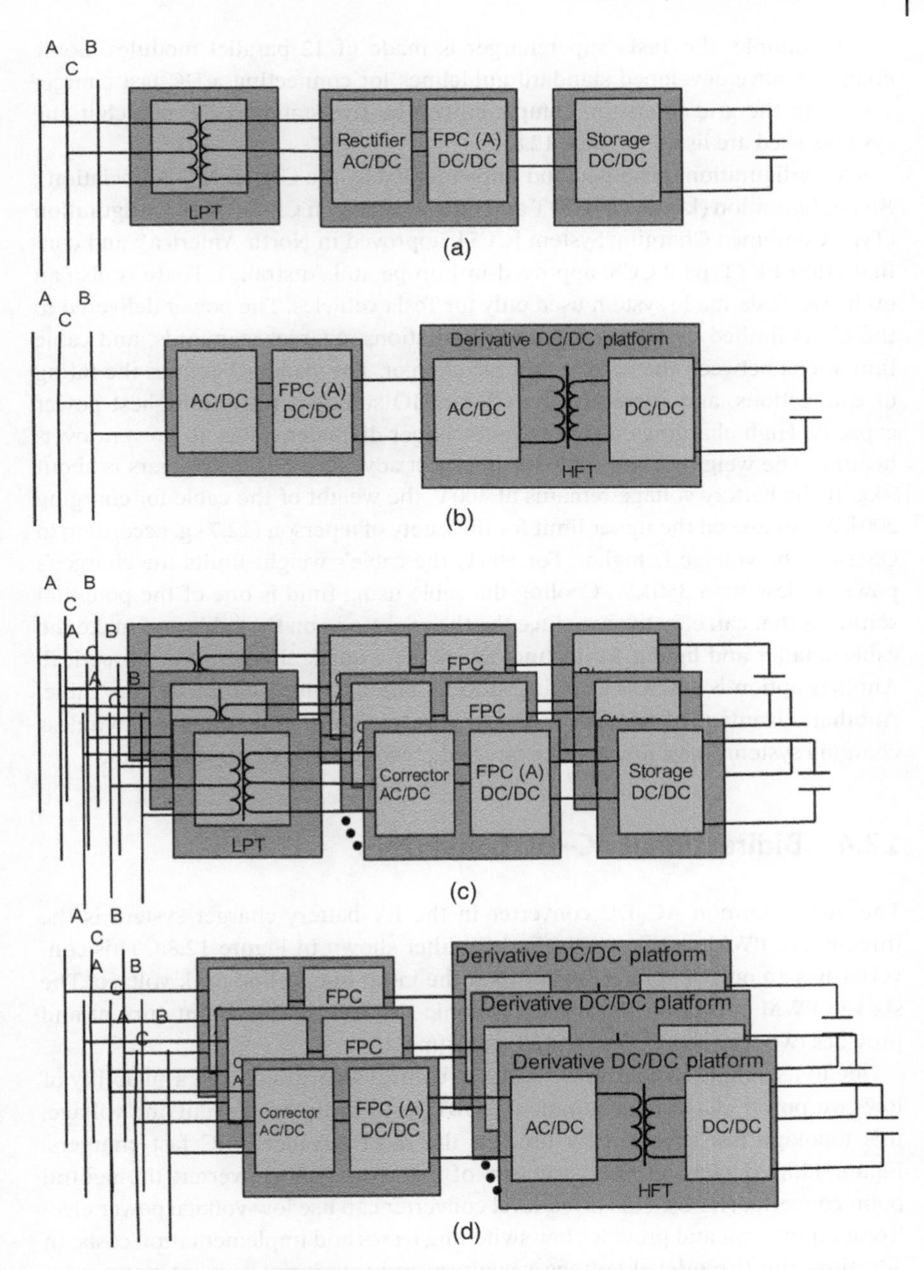

Figure 12.7 Typical DC fast charger power converter diagrams. (a) A single module charger with a non-isolated DC–DC converter; (b) a single module charger with an isolated DC–DC converter; (c) several parallel modules shown in (a); (d) several parallel modules shown in (b).

For example, the Tesla supercharger is made of 12 parallel modules. Some countries have developed standard guidelines for connecting a DC fast charger system to the grid to ensure compatibility. The five standard DC fast charging systems used are listed in Table 12.8.

AA configuration (proposed and implemented by the CHAdeMO Association), BB configuration (known as GB/T and only available in China), EE configuration (Type Combined Charging System [CCS], approved in North America), and configuration FF (Type 2 CCS, approved in Europe and Australia). There is also an exclusive Tesla-made system used only for Tesla vehicles. The power delivered to the EV is limited by battery charging limitations, charger standards, and cable limitations between the vehicle and the charger. The standard defines the rating of connections, and currently, the CHAdeMO standard has the highest power capacity. High charging current requires larger-diameter cables to prevent overheating. The weight of the cable for the most advanced 50 kW chargers is about 9 kg. If the battery voltage remains at 400 V, the weight of the cable for charging 200 kW can exceed the upper limit for the safety of a person (22.7 kg, according to OSHA). The voltage is higher. For 800 V, the cable's weight limits the charger's power to less than 350 kW. Cooling the cable using fluid is one of the potential solutions that can effectively reduce the thermal stress on the cable and make the cable smaller and lighter for technology. Make a quick charger (XFC) practical. Another option is wireless charging at XFC stations, which eliminates the cable. Another advantage of wireless charging is galvanic isolation. However, wireless charging systems have lower efficiency and power density than wired ones.

12.4 Bidirectional AC–DC Converters

The most common AC–DC converter in the EV battery charger system is the three-phase PWM rectifier with the LCL filter shown in Figure 12.8a. This converter has an output voltage higher than the input line-to-line peak voltage. The six-key PWM converter has a low harmonic spectrum in the input current and provides two-way power flow and PF adjustment.

Due to its simple structure, well-known control algorithms, and availability of low-cost power electronic equipment (IGBT) with sufficient current and voltage, this topology has been widely used in the most advanced DC fast chargers. Figure 12.8b depicts another structure of the boost-type converter: the neutral point converter (NPC). This three-level converter can use low-voltage power electronic equipment and provides low switching losses and implementation costs. In addition, the three-level voltage waveform reduces input current harmonics and dv/dt. As an example, in reference [16], in an electric car charger rated at 30 kW using the NPC front-end converter, the leakage inductance of the input

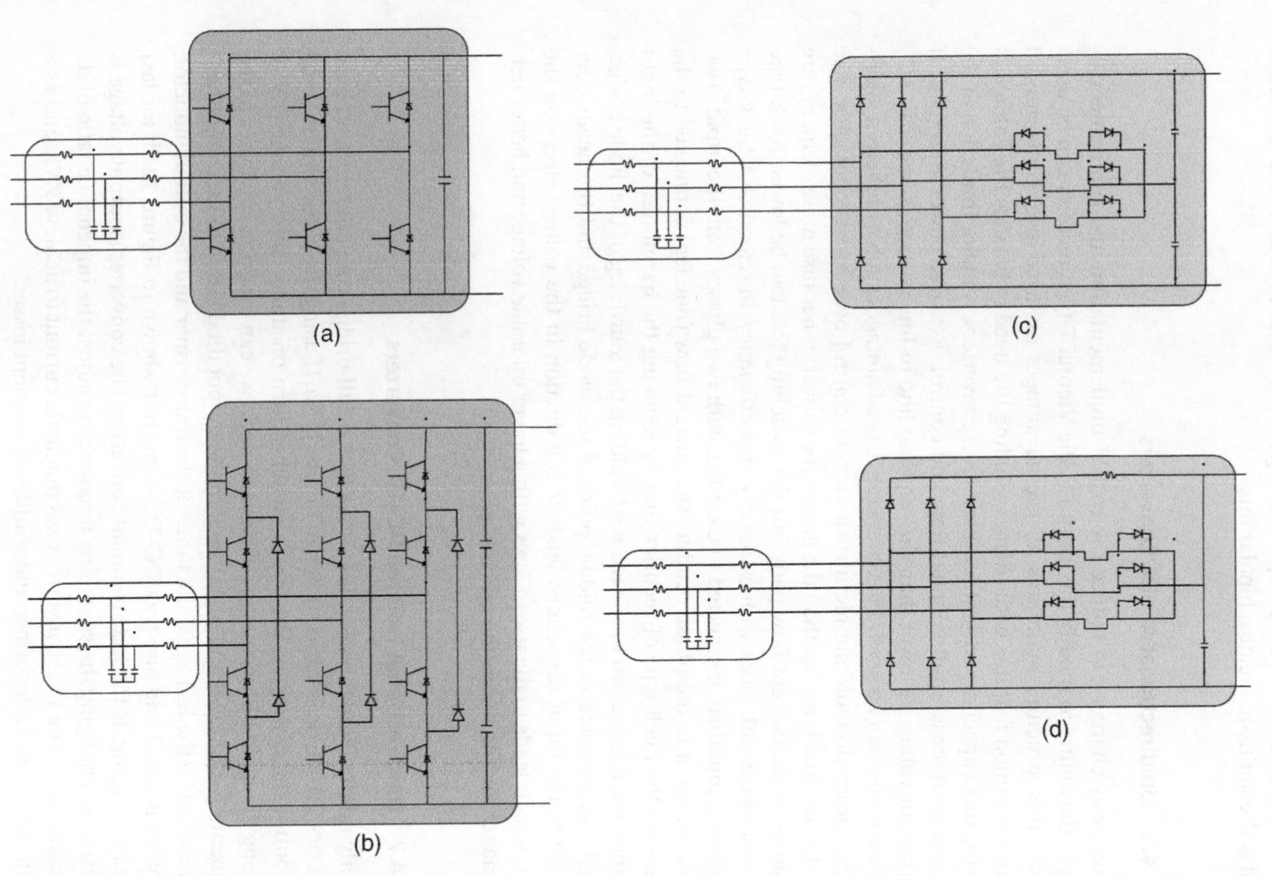

Figure 12.8 AC–DC converter structures for DC fast chargers. (a) Three-phase PWM rectifier; (b) neutral point-clamped rectifier; (c) Vienna rectifier; (d) Buck type rectifier.

transformer, acting as an AC side filter, reduces the harmonic spectrum of the input current. Another advantage of using the NPC converter as an AC–DC front-end is the ability to create a bipolar bus.

12.4.1 Unidirectional AC–DC Converters

In one-way chargers, to reduce the cost of implementation and simultaneously create the ability to modify the input PF, the Vienna T-type rectifier can be used. While this structure maintains all the advantages of three-level converters, it shares common features with them, including the need to balance the voltage of the DC link capacitors with other three-level converters. A major limitation of the Vienna rectifier is unidirectional flow and control. Reactive power is limited. If the output voltage is lower than the required line-to-line voltage, a one-way AC–DC converter can be used. This converter has advantages such as inherent short-circuit protection and simple inrush current control over boost topologies. The conduction loss is higher than the boost type converter because more elements are in series in the current flow path, but the switching loss can be lower. Buck-type converters can still work with extremely high efficiency. In reference, a buck-type rectifier is modified, connected to each leg, with two phases from the input. Two legs connected to the phase conduct the current flow (one leg) connected to the phase for the buck type of rectifier, and by reducing the harmonics of the input current, the conduction loss is reduced. Adding the fourth leg of the diode bridge, which is connected to the middle point of the diode bridge and the heads connected to the input capacitors, leads to a reduction in the voltage stress on the keys, which leads to the use of keys with a lower tolerable voltage and better performance and increases the efficiency.

12.4.2 Unidirectional Isolated DC–DC Converters

If only unidirectional power flow is needed, a full-bridge converter with a phase shift (PSFB) can be used, as shown in Figure 12.9a. If using the phase shift method to control the converter, the power switches turn on under zero voltage (ZVS). Among the main disadvantages of this topology, we can mention key losses at the moment of shutdown and high losses in the output diodes due to LCL amplification caused by the leakage inductance of the transformer and the output inductor. Another isolated and one-way DC–DC converter, shown in Figure 12.9b for fast battery charging, is the LLC resonant converter. The converter's output voltage is adjusted by changing the switching frequency to adjust the impedance ratio of its resonant tank. The LLC converter uses magnetic current to achieve ZVS, and as a result, it has low transformer and shutdown moment losses.

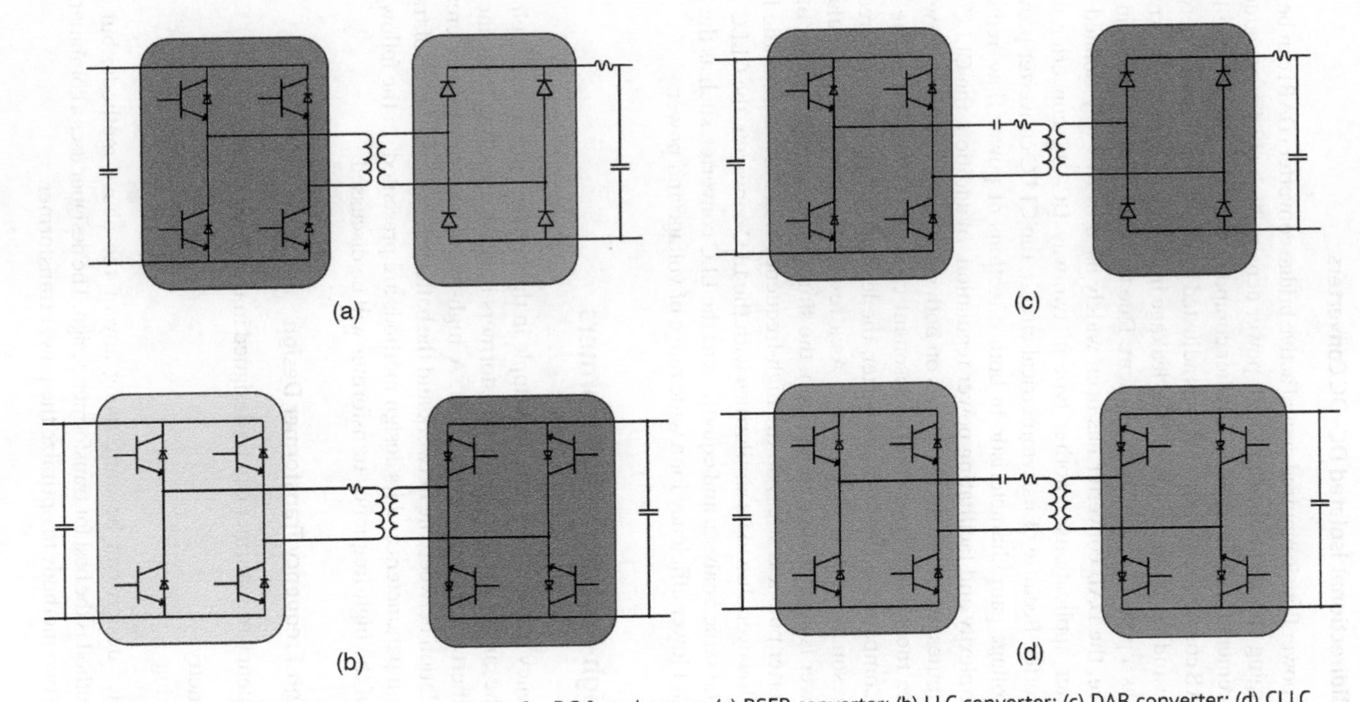

Figure 12.9 Isolated DC–DC converter structures for DC fast chargers. (a) PSFB converter; (b) LLC converter; (c) DAB converter; (d) CLLC converter.

12.4.3 Bidirectional Isolated DC–DC Converters

If two-way power flow is required, a dual active bridge converter (DAB) can be used for EV charging operation due to its high-power density, high efficiency, reduction and enhancement capability, and small filter components as shown in Figure 12.9c.

In the DAB converter, the power flow is adjusted by changing the phase between the primary and secondary voltages. The leakage inductance of the transformer is controlled as a power transmission element. Due to its simple structure and ZVS performance, the DAB converter has been widely used in two-way isolated DC–DC converter applications. Another type of two-way DC–DC converter is the CLLC converter. Because of its symmetrical circuit, the CLLC converter provides the same voltage gain characteristic in both directions of power flow, reducing control complexity and facilitating power regulation. In addition, the CLLC converter distributes two resonant capacitors on both sides of the transformer, which helps reduce the voltage stress on the resonant capacitor compared to the LLC converter. Compared to the DAB converter, the leakage inductance required for the CLLC resonant tank is much lower. As a result, the converter's circulating reactive power is also smaller. In addition, the sinusoidal resonant current of the CLLC converter puts less stress on the high-frequency transformer than the DAB converter. However, due to its similarities with the LLC converter, the CLLC converter has the same analyses and equations as the LLC converter, such as the ZVS condition and lower efficiency for a wide range of voltage and power.

12.5 High-Frequency Transformers

High-frequency transformers have less ripple in the voltage output, and their use can make the circuit smaller. These transformers have an operating frequency of several kilohertz and have ferrite cores. A high-frequency transformer creates galvanic isolation between the network and the batteries. The calculation of transformer design parameters and its design methods are presented. In the following, the design of the high-frequency transformer will be discussed.

12.5.1 High-Frequency Transformer Design

A high-frequency transformer can be designed in two diverse ways:

- Core geometry
- Surface thing

Transformer design can be done using any of the above methods, but the combined method is the best for transformer design. The designer uses a combination of the above two methods to optimize the power transformer.

The following important items are used in the design of high-frequency transformers:

- The number of winding turns and their rounding method
- Copper depletion
- Core loss
- Optimal determination of flux density with maximum efficiency
- Adjacency and skin phenomena
- Minimum dimensions of the core for the transformer
- Methods 1–6: Surface product

In this method, the core dimensions required for the transformer depend on the product of the core's effective cross-sectional area and the winding window's area.

12.5.2 Core Geometry Method

Core losses and winding losses are important challenges in transformer design. The total loss of the transformer depends on the wire diameter, flux density, frequency, core material, and window factor. By optimally choosing these parameters, core size and losses can be reduced. The core loss depends on flux density. To choose the optimal core, the flux density should be optimized. Also, to optimize the core, you can optimize the frequency. The optimal frequency is used because the most significant losses are switching losses. Winding losses include proximity effect and skin effect losses. To reduce the proximity effect, the sandwich twisting method is used. Conductors with a cross-sectional area in the skin depth range are used to limit skin effect losses. The winding loss can be reduced by optimally choosing the window coefficient, which is the winding loss coefficient of this parameter. In the design of the transformer, the leakage inductance is assumed to be large. Core and copper losses are considered in the design. The design method will be explained based on the constant geometric criterion of the core (Kgfe).

12.5.3 Core Losses

In choosing a core with different compositions and alloys, there is a compromise between core loss and saturation magnetic flux density. Choosing cores with high BSAT leads to many losses, but it reduces the price, volume, and size. Ferrites have a lower B_{sat} $(T > B_{sat} > 0.5\ T\ 0.2)$ than other materials. Due to the low resistance of the ferrite core, the eddy losses are less, and they have good efficiency in the frequency range between tens of kilohertz and one megahertz. Using Eq. (12.5) for a certain frequency, the Pfe core loss is approximated.

$$P_{fe} = K_{fe}\left(\Delta B\right)^{\beta} A_c\, L_m \tag{12.5}$$

where L_m is the average magnetic length of the core and A_c is the cross-section of the core. A typical value of β in ferrite cores is $2.6 > \beta > 28$. In Eq. (12.5), the harmonic effects of the applied waveform on the transformer are ignored, and the sinusoidal waveform is assumed.

12.6 Examine Some of the Charger Examples Provided in the References

Electric vehicles have had various charger circuits since the beginning of their production. Chargers can be classified based on the location and type of connection, the amount of power, and the direction of power flow. The most common method for a fast battery charger is to choose the appropriate AC and DC voltage levels. In one-way chargers, power flows only from the network side to the batteries, and in two-way chargers, it is possible to flow power in both directions, i.e. from the network to the vehicle (G2V) and from the vehicle to the network (V2G). In order for the connection of the charger to the network not to cause power quality problems, the current drawn or injected into the network in V2G and G2V modes must have a sinusoidal waveform and be in phase with the network voltage. In [17], a single-stage structure for off-board battery charging as shown in Figure 12.10. This structure can directly charge the battery without a DC–DC converter. This structure, based on a thyristor rectifier, is a cost-effective solution. It is

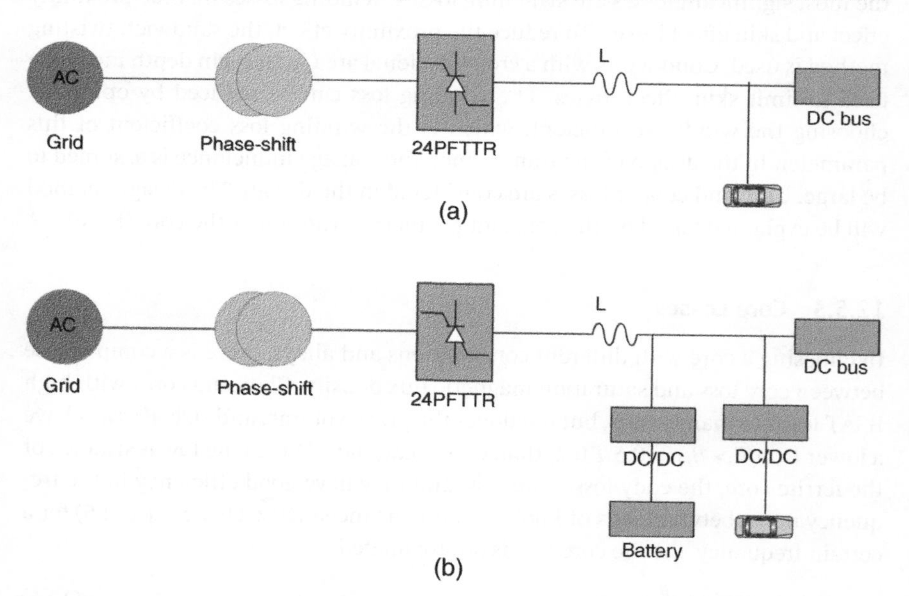

Figure 12.10 Off-board strategy for the battery charger. (a) Independent battery charger at the charging station; (b) battery charger with energy storage.

recommended to provide a DC terminal for the charging station, especially in high-power applications. Due to the low-frequency operation, switching losses are reduced. Using transformers has increased the volume of the structure and reduced its efficiency.

In reference [18], a new structure for a two-stage battery charger for EVs is proposed based on a two-stage converter, as shown in Figure 12.11. The first stage (the AC–DC converter) is used as an interface between the power grid and the DC link and consists of two parallel three-phase full-bridge converters. The second stage is the DC–DC section, which serves as an interface between the DC link and the battery and consists of a two-way, three-level voltage converter. In this converter, it is possible to achieve sinusoidal network current and unit PF even with harmonic network voltage.

Sixteen power switches are used in this structure, which complicates the control and increases switching losses, as well as the construction cost. In reference [19], a torque removal strategy for the 1IBC topology based on permanent magnet six-phase synchronous machines is presented. The proposed structure in this reference can be seen in Figure 12.12. This strategy does not interfere with charging or discharging the battery, and there is no need for additional components in the circuit. Fourteen power switches are used in this structure, which increases switching losses. The control method for this structure is complex. In this structure, using a six-phase motor complicates the control and requires additional components, such as the rotor position sensor in the control circuit, which increases the manufacturing cost.

In reference [20], a developed EV battery charger consists of a front-end AC–DC converter and a back-end DC–DC converter, separated by a capacitive DC link as shown in Figure 12.13. EV battery charge control (also known as "power control" or "P-Control") and network voltage control (V-Control) are performed by DC–DC converters and AC–DC converters, respectively. Harmonic currents on the grid side are regulated by the AC–DC converter and control systems provided within the standard limits.

In reference [21], a high-efficiency isolated battery charger is proposed for light electric vehicles, as shown in Figure 12.13. Eliminating the input diode bridge in the proposed structure has reduced conduction losses. The proposed control algorithm enables the charger to modify the PF and adjust the output voltage. With these capabilities, the charger structure proposed in this reference has high efficiency and good power quality. Therefore, this charger with the proposed control algorithm is an effective solution for charging EV batteries. This charger is used for light loads.

In reference [22], a control loop (SMC) for ILTP converters in separate phases of on-board chargers is proposed as shown in Figure 12.14. The main objectives of the proposed SMC are to improve the converter's PF and ensure the converter's output voltage regulation under rapid load fluctuations. This structure is designed for low power and long charging time.

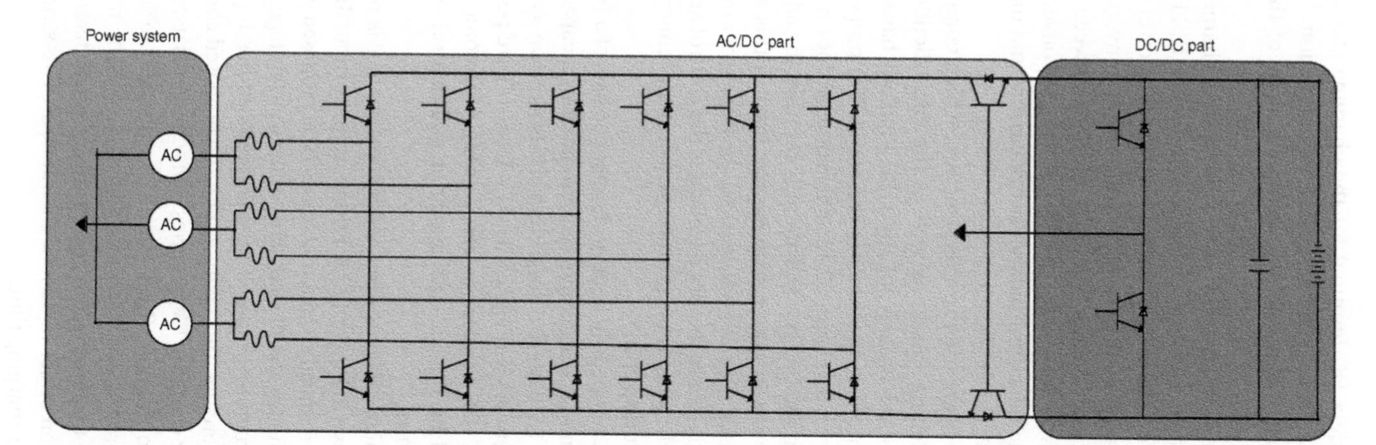

Figure 12.11 Structure of a three-phase fast battery charger for electric vehicles.

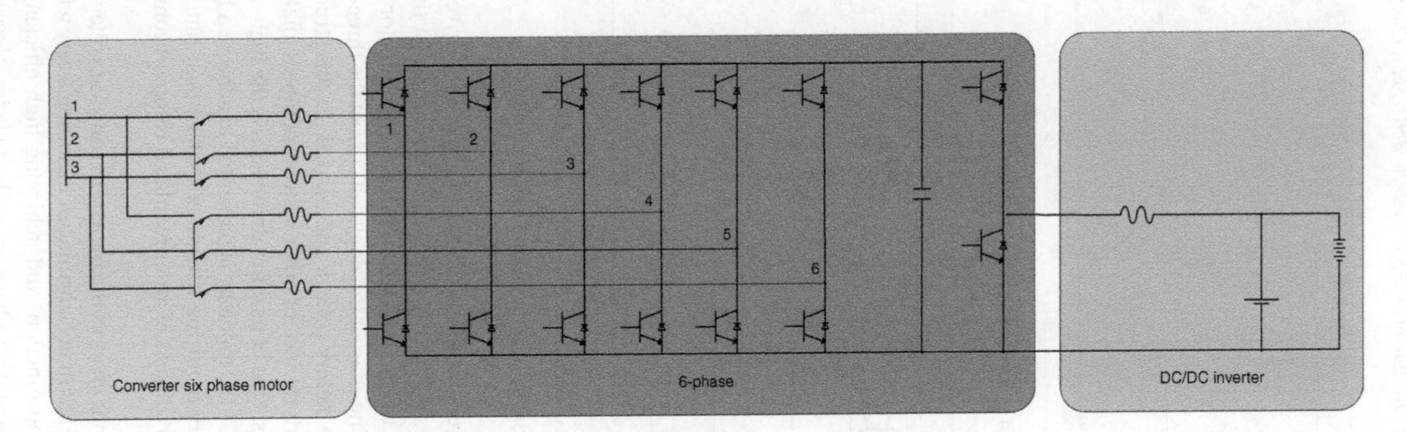

Figure 12.12 IBC-type battery charger structure.

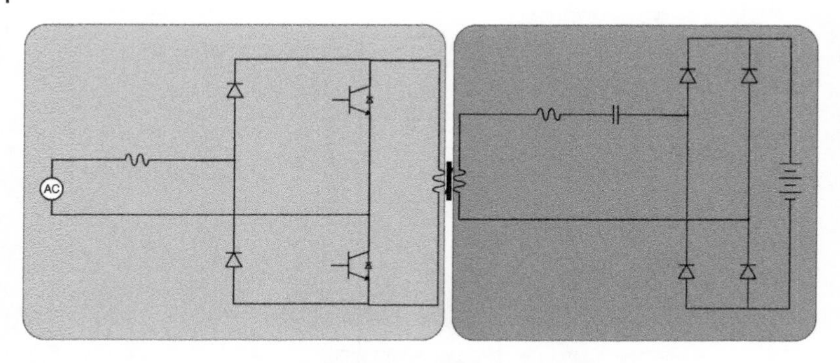

Figure 12.13 Structure of an isolated battery charger.

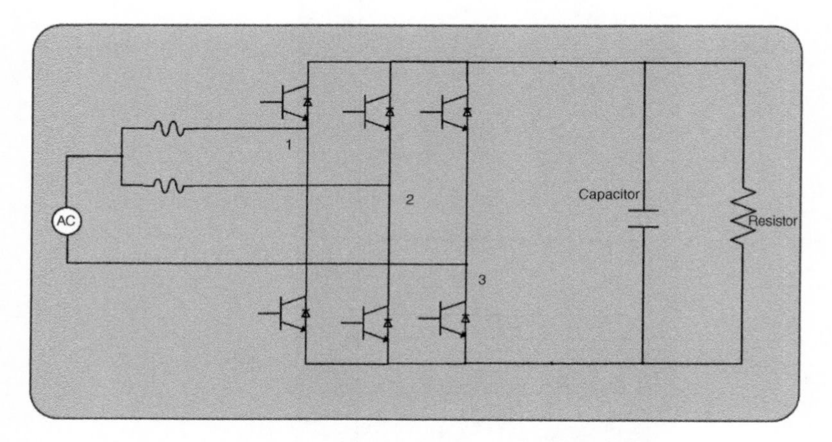

Figure 12.14 The basic circuit structure of the totem pole PFC.

In reference [23], a three-phase on-board charger with the drive of a plug-in electric vehicle (PEV) is presented. It is designed by connecting a three-phase power electronic interface and using the drive motor as a DC inductor connected to the charger (Figure 12.15). The charger topology can correct the three-phase power factor (PFC) and adjust the battery voltage and current. In this structure, if the parts fail, it may affect the performance of other system members, making it difficult to access the damaged parts. One of its other drawbacks is the high construction cost.

In reference [24], a DC–DC topology is presented to achieve a high efficiency of 97% for EV fast chargers requiring a wide output voltage range. The proposed topology in Figure 12.16 consists of an unregulated resonant converter and a non-isolated buck converter.

The step-down converter charges the battery in constant voltage or current control mode. This structure has many advantages, including a very wide voltage range, no zero-voltage switching function, and achieving high efficiency. The modular structure and the use of many keys have led to the switching losses of this topology

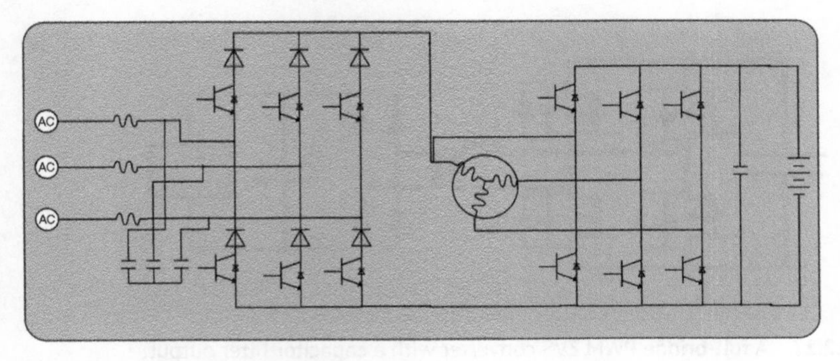

Figure 12.15 Proposed structure for PEV.

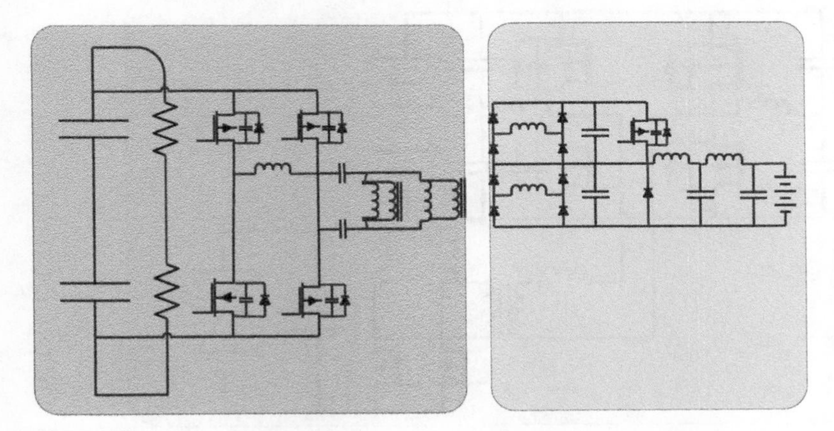

Figure 12.16 Proposed resonant converter and a non-isolated buck converter.

and increased its construction cost. In reference [25], a full-bridge converter with modulation (PWM) and a capacitive output filter is presented (Figure 12.17). This research aims to achieve high efficiency and low cost by minimizing the charger size, charging time, and electricity cost. In this structure, the rectifier diodes are off at zero current, and as a result, no clamp circuit is needed at the output. This converter works in continuous and discrete current modes. The power of this converter is low due to the low output voltage (400 V) and its use in hybrid cars.

In reference [26], an adaptive load control approach is presented to optimally control the reactive current required for zero voltage switching (ZVS). The DC–DC converter is a battery charger for an EV (Figure 12.18). Since this structure requires a wide range of load changes, the converter must be able to maintain ZVS from full-load to no-load conditions. The converter uses an asymmetric auxiliary circuit to provide reactive current for the semiconductor switches of the full-bridge converter, which ensures ZVS when the switch is turned on. A high circulating current at low power levels reduces efficiency.

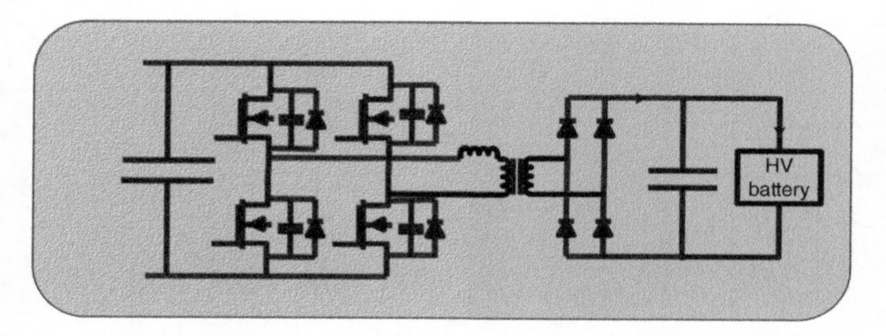

Figure 12.17 A full-bridge PWM ZVS converter with a capacitor filter output.

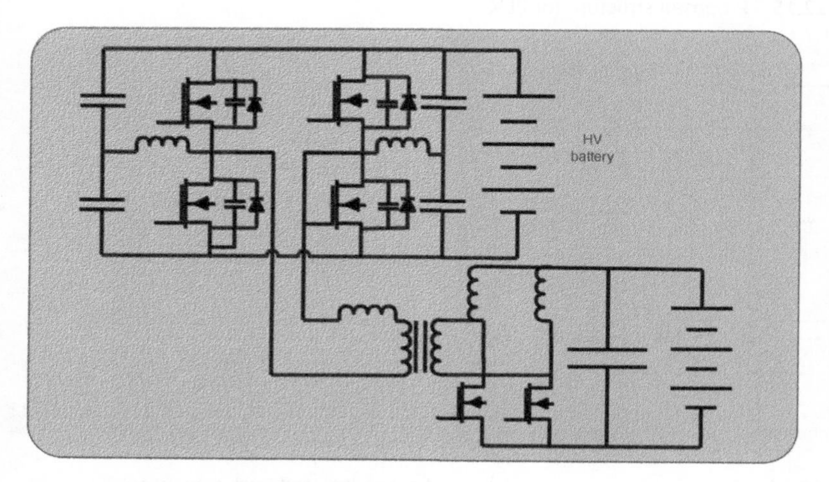

Figure 12.18 Full-bridge ZVS converter.

12.7 Conclusion

EV battery chargers consist of two parts. AC–DC converters are used in the first part, and DC–DC converters are used in the second part. In this research, lithium-ion battery charging algorithms were investigated, and finally, they were compared. Further, the power circuits of chargers and some examples of references were reviewed, and some advantages and disadvantages were mentioned. In AC–DC converters, unit PF and voltage rectification are crucial factors to check. The weight, size, and cost of making these converters are other crucial factors compared with other converters. Every converter has some problems. One of the key factors to consider in DC–DC converters is the output current ripple. The main problems in some converters were the high voltage stress on the switches, low efficiency, high manufacturing cost, inappropriate current for charging the battery, etc. Finally, this article reviewed lithium-ion batteries, and their performance was compared with other batteries. The control systems of lithium-ion batteries were also investigated.

References

1 Zand, M., Nasab, M.A., Hatami, A., Kargar, M., and Chamorro, H.R. (2020). Using adaptive fuzzy logic for intelligent energy management in hybrid vehicles. *2020 28th Iranian Conference on Electrical Engineering (ICEE)*, pp. 1–7. IEEE. 10.1109/ICEE50131.2020.9260941. IEEE Index.

2 Ahmadi-Nezamabad, H., Zand, M., Alizadeh, A. et al. (2019). Multi-objective optimization based robust scheduling of electric vehicles aggregator. *Sustainable Cities and Society* 47: 101494.

3 Zand, M., Nasab, M.A., Sanjeevikumar, P. et al. (2020). Energy management strategy for solid-state transformer-based solar charging station for electric vehicles in smart grids. *IET Renewable Power Generation* 14 (18): 3843–3852.

4 Ghasemi, M., Akbari, E., Zand, M. et al. (2019). An efficient modified HPSO-TVAC-based dynamic economic dispatch of generating units. *Electric Power Components and Systems* 47 (19–20): 1826–1840.

5 Nasri, S., Nowdeh, S.A., Davoudkhani, I.F. et al. (2021). Maximum power point tracking of photovoltaic renewable energy system using a new method based on turbulent flow of water-based optimization (TFWO) under partial shading conditions. In: *Fundamentals and Innovations in Solar Energy*, 285–310. Singapore: Springer.

6 Rohani, A., Joorabian, M., Abasi, M., and Zand, M. (2019). Three-phase amplitude adaptive notch filter control design of DSTATCOM under unbalanced/distorted utility voltage conditions. *Journal of Intelligent & Fuzzy Systems* 37 (1): 847–865.

7 Zand, M., Nasab, M.A., Neghabi, O. et al. (2019). Fault locating transmission lines with thyristor-controlled series capacitors by fuzzy logic method. In: *2020 14th International Conference on Protection and Automation of Power Systems (IPAPS)*, 62–70. IEEE.

8 Zand, Z., Hayati, M., and Karimi, G. (2020). "Short-channel effects improvement of carbon nanotube field effect transistors," 2020 28th Iranian Conference on Electrical Engineering (ICEE), Tabriz, Iran, 2020, pp. 1–6, doi: https://doi.org/10.1109/ICEE50131.2020.9260850.

9 Tightiz, L., Nasab, M.A., Yang, H., and Addeh, A. (2020). An intelligent system based on optimized ANFIS and association rules for power transformer fault diagnosis. *ISA Transactions* 103: 63–74.

10 Zand, M., Neghabi, O., Nasab, M.A. et al. (2020). A hybrid scheme for fault locating in transmission lines compensated by the TCSC. In: *2020 15th International Conference on Protection and Automation of Power Systems (IPAPS)*, 130–135. IEEE.

11 Nasab, A., Morteza, M.Z., Eskandari, M. et al. (2021). Optimal planning of electrical appliance of residential units in a smart home network using cloud services. *Smart Cities* 4 (3): 1173–1195.

12 Sanjeevikumar, P., Zand, M., Nasab, M.A. et al. (2021). Spider community optimization algorithm to determine UPFC optimal size and location for improve dynamic stability. In: *2021 IEEE 12th Energy Conversion Congress & Exposition-Asia (ECCE-Asia)*, 2318–2323. IEEE.

13 Azimi Nasab, M., Zand, M., Eskandari, M. et al. (2021). Optimal planning of electrical appliance of residential units in a smart home network using cloud services. *Smart Cities* 4 (3): 1173–1195.

14 Nasab, M.A., Zand, M., Padmanaban, S. et al. (2022). An efficient, robust optimization model for the unit commitment considering renewable uncertainty and pumped-storage hydropower. *Computers and Electrical Engineering* 100: 107846.

15 Azimi Nasab, M., Zand, M., Padmanaban, S., and Khan, B. (2021). Simultaneous long-term planning of flexible electric vehicle photovoltaic charging stations in terms of load response and technical and economic indicators. *World Electric Vehicle Journal* 12 (4): 190.

16 Zand, M., Nasab, M.A., Padmanaban, S., and Khoobani, M. (2022). Big data for SMART sensor and intelligent electronic devices–building application. In: *Smart Buildings Digitalization*, 11–28. CRC Press.

17 Padmanaban, S., Khalili, M., Nasab, M.A. et al. (2022). Determination of power transformers health index using parameters affecting the transformer's life. *IETE Journal of Research* 1–22.

18 Sanjeevikumar, P., Samavat, T., Nasab, M.A. et al. (2022). Machine learning-based hybrid demand-side controller for renewable energy management. In: *Sustainable Developments by Artificial Intelligence and Machine Learning for Renewable Energies*, 291–307. Elsevier.

19 Khalili, M., Ali Dashtaki, M., Nasab, M.A. et al. (2022). Optimal instantaneous prediction of voltage instability due to transient faults in power networks, taking into account the dynamic effect of generators. *Cogent Engineering* 9 (1): 2072568.

20 Asadi, A.H.K., Jahangiri, A., Zand, M. et al. (2022). Optimal design of high density HTS-SMES step-shaped cross-sectional solenoid to mechanical stress reduction. In: *2022 International Conference on Protection and Automation of Power Systems (IPAPS)*, vol. 16, 1–6. IEEE.

21 Nasab, M.A., Zand, M., Hatami, A. et al. (2022). A hybrid scheme for fault locating for transmission lines with TCSC. In: *2022 International Conference on Protection and Automation of Power Systems (IPAPS)*, vol. 16, 1–10. IEEE.

22 Dashtaki, A.A., Khaki, M., Zand, M. et al. (2022). A day ahead electrical appliance planning of residential units in a smart home network using ITS-BF algorithm. *International Transactions on Electrical Energy Systems*.

23 Koç, Y., Birbir, Y., and Bodur, H. (2022). Non-isolated high step-up DC/DC converters–an overview. *Alexandria Engineering Journal* 61 (2): 1091–1132.

24 Dhananjaya, M. and Pattnaik, S. (2022). Review on multi-port DC–DC converters. *IETE Technical Review* 39 (3): 586–599.

25 Verma, D., Nema, S., Agrawal, R. et al. (2022). A different approach for maximum power point tracking (MPPT) using impedance matching through non-isolated DC-DC converters in solar photovoltaic systems. *Electronics* 11 (7): 1053.

26 Rajesh, P., Shajin, F.H., and Kommula, B.N. (2022). An efficient integration and control approach to increase the conversion efficiency of high-current low-voltage DC/DC converter. *Energy Systems* 13 (4): 939–958.

Index

Fast-Charging Infrastructure for Electric and Hybrid Electric Vehicles: Methods for Large-Scale Penetration into Electric Distribution Networks, First Edition. Sivaraman Palanisamy, Sharmeela Chenniappan, and Sanjeevikumar Padmanaban.

Published 2023 by John Wiley & Sons, Inc.

Printed and bound by CPI Group (UK) Ltd, Croydon, CR0 4YY

16/04/2025

14658598-0002